EXPLORATIONS IN CALCULUS WITH A COMPUTER ALGEBRA SYSTEM

International Series in Pure and Applied Mathematics

Ahlfors: *Complex Analysis*
Bender and Orszag: *Advanced Mathematical Methods for Scientists and Engineers*
Boas: *Invitation to Complex Analysis*
Buck: *Advanced Calculus*
Colton: *Partial Differential Equations*
Conte and deBoor: *Elementary Numerical Analysis: An Algorithmic Approach*
Edelstein-Keshet: *Mathematical Models in Biology*
Goldberg: *Matrix Theory with Applications*
Hill: *Experiments in Computational Matrix Algebra*
Lewin and Lewin: *An Introduction to Mathematical Analysis*
Morash: *Bridge to Abstract Mathematics*
Parzynski and Zipse: *Introduction to Mathematical Analysis*
Pinsky: *Partial Differential Equations and Boundary Value Problems with Applications*
Pinter: *A Book of Abstract Algebra*
Ralston and Rabinowitz: *A First Course in Numerical Analysis*
Ritger and Rose: *Differential Equations with Applications*
Rudin: *Functional Analysis*
Rudin: *Principles of Mathematical Analysis*
Rudin: *Real and Complex Analysis*
Small and Hosack: *Calculus: An Integrated Approach*
Small and Hosack: *Explorations in Calculus with a Computer Algebra System*
Vanden Eynden: *Elementary Number Theory*
Simmons: *Differential Equations with Applications and Historical Notes*
Walker: *Introduction to Abstract Algebra*

Churchill–Brown Series

Complex Variables and Applications
Fourier Series and Boundary Value Problems
Operational Mathematics

Also available from McGraw-Hill

Schaum's Outline Series in Mathematics and Statistics

Most Outlines include basic theory, definitions, and hundreds of solved problems
and supplementary problems with answers.

TITLES ON THE CURRENT LIST INCLUDE:

Advanced Calculus
Advanced Mathematics
Analytic Geometry
Beginning Calculus
Boolean Algebra
Calculus, 3d edition
Calculus of Finite Differences &
   Difference Equations
College Algebra
Complex Variables
Descriptive Geometry
Differential Equations
Differential Geometry
Discrete Math
Elementary Algebra
Review of Elementary Mathematics
Essential Computer Math
Finite Mathematics
First Year College Mathematics
Fourier Analysis
General Topology

Geometry, 2d edition
Group Theory
Laplace Transforms
Linear Algebra
Mathematical Handbook of Formulas & Tables
Matrices
Matrix Operations
Modern Algebra
Modern Elementary Algebra
Modern Introductory Differential Equations
Numerical Analysis, 2d edition
Partial Differential Equations
Probability
Probability & Statistics
Projective Geometry
Real Variables
Set Theory & Related Topics
Statistics, 2d edition
Tensor Calculus
Trigonometry, 2d edition
Vector Analysis

Schaum's Solved Problem Books

Each title in this series is a complete and expert source of solved problems containing
thousands of problems with worked out solutions

TITLES ON THE CURRENT LIST INCLUDE:

3000 Solved Problems in Calculus
2500 Solved Problems in College Algebra and Trigonometry
2500 Solved Problems in Differential Equations
3000 Solved Problems in Linear Algebra
3000 Solved Problems in Numerical Anaylsis
3000 Solved Problems in Precalulus
2000 Solved Problems in Statistics

Available at your College Bookstore. A complete list of Schaum titles may be obtained by
writing to: Schaum Division
         McGraw-Hill, Inc,
         Princeton Road S-1
         Hightstown, NJ 08520

EXPLORATIONS IN CALCULUS WITH A COMPUTER ALGEBRA SYSTEM

DONALD B. SMALL

Colby College

JOHN M. HOSACK

University of the South Pacific

McGraw-Hill, Inc.

New York   St. Louis   San Francisco   Auckland   Bogotá   Caracas
Hamburg   Lisbon   London   Madrid   Mexico   Milan   Montreal   New Delhi
Paris   San Juan   São Paulo   Singapore   Sydney   Tokyo   Toronto

Explorations in Calculus with a Computer Algebra System

Calculus T/L is a trademark of Wadsworth, Inc.
Derive and **muMath** are registered trademarks of Soft Warehouse, Inc.
IBM is a registered trademark of International Business Machines Incorporated
Macintosh is a registered trademark of Apple Computers, Inc.
MACSYMA is a trademark of Symbolics, Inc.
Maple is a registered trademark of the University of Waterloo.
Mathematica is a trademark of Wolfman Research, Inc.
Microsoft and **MS-DOS** are registered trademarks of MicroSoft Corporation
PostScript is a registered trademark of Adobe Systems, Inc.
SMP is a trademark of Innosoft International, Inc.
Theorist is a registered trademark of Allan Bonadio Associates/Prescience Corporations
UNIX is a trademark of AT&T.
VMS is a trademark of the Digital Equipment Corporation.

2 3 4 5 6 7 8 9 0 DOC DOC 9 5 4 3 2 1

ISBN 0-07-058267-X

The editor was Richard Wallis;
the production supervisor was Denise L. Puryear.
R. R. Donnelley & Sons Company was printer and binder.

Library of Congress Catalog Card Number: 90-62227

It is a pleasure to dedicate my work on this book to the "kids"

   Jim & Peg   *Jan & Jeff*   *Bob & Donna*

D.B.S.

It is an honor to dedicate my work on this book to my parents

             Mr. and Mrs. Gayle Hosack

J.M.H.

Contents

Preface

Traditionally the focus of introductory calculus classes has been on computation. The emphasis of classes, homework, and examinations has been on the student's ability to carry out certain algorithms: for differentiation, integration, limit, and series computations. With the widespread availability of Computer Algebra Systems (CASs, see below) which can carry out these algorithms better than people, this emphasis has become anachronistic. The emphasis should be on concepts, understanding, and applications. The CASs can handle the routine computations while the student directs the analysis. This allows more examples and more complex examples to be considered. The student can focus on concepts rather than on computational details. With a CAS handling the tedious computations, students can explore concepts by looking at examples and searching for patterns. CASs allow for the integration of graphical, symbolic, and numerical methods in analyzing problems.

The purpose of the explorations in this text is to encourage students to use computer algebra systems as tools to further their understanding of concepts and applications.

There are three parts to this text. The first part is a brief introduction to computer algebra systems and motivation for their use. The second part consists of relatively independent sections that apply computer algebra systems to calculus. Each section has a description of the required knowledge to use the section, discussion for the section, worked examples (the ends of which are indicated by the symbol $\triangle$), and exercises. Some of the sections cover standard topics in calculus; others provide brief introductions to new topics. Because of the independence of the topics, there are many possible uses of this text. The third part consists of introductions to the use of several available computer algebra systems.

To the Instructor

It would be a mistake to incorporate CASs into courses primarily as exercise solvers while continuing our present orientation towards carrying out algorithmic computations. Our goals, expectations, assignments, and classroom instruction need to change to maximize the opportunities afforded by modern technology.

We briefly describe six areas where CASs can effect major changes in the under-graduate program (for more discussion see [11]).

Student perception of what is important in mathematics

Students normally measure the importance of an activity by the amount of time spent on it and the proportion of the examinations allotted to it. Since most of a student's effort on both homework and tests is devoted to algorithmic com-putation, it is not surprising that students view mathematics as a collection of formulas (to be memorized), and "to do mathematics" is to compute.

Several things can happen when routine algorithmic computations are rele-gated to a machine. Most important of all, time is made available for concen-trating on concepts, motivation, applications, and ramifications. Computational mistakes are eliminated. Self confidence will develop from understanding and applying concepts rather than from being a computational robot. Computation will be viewed as a means rather than as an end.

CASs can be effective tools for convincing students that the appropriate focus should be on concepts and processes rather than on mastery of algorithms.

The role of approximation and error bound analysis

Approximation and error bound analysis is at the heart of applied mathematics as well as being the backbone of analysis. However, this analysis is only slightly touched on in elementary calculus because of the extensive computations that are often involved. Thus, in this situation, the algebra is the restricting influence on both the type and level of inquiry. In a more general sense, algebraic limitations are a prime cause for the dependence of our elementary courses on closed form solutions. This, in turn, eliminates or restricts modeling in elementary courses to artificially constructed examples and leads students to question the value or applicability of what they are doing.

A CAS can be an effective tool in shifting the emphasis from closed form to open ended problems. In particular, numerical integration can be developed as the "norm" or "standard" with closed form integration being considered as a special case. The availability of CASs can open the door to modeling in our introductory classes, so that our students can experience more realistic applica-tions.

The role of graphical analysis

CASs are interchanging the order of graphing and analysis. Graphing has tradi-tionally been an application of analysis. However, graphics packages that allow for zooming, digitizing, root finding, etc. will be used to "lead" the analysis. For example, graphics greatly change the treatment of finding extremal values. Also solving equalities and inequalities is easily accomplished by graphically finding the zeros of an appropriate function. The (in)famous exercises of finding a δ

given an ϵ will be done by reading zeros off a plot of an appropriate function. Graphics free the user from the severe algebraic limitations of factoring. In fact, transforming a problem into one of finding the zeros of a function (which will be done graphically) may become one of the most used techniques in calculus.

Development of problem solving skills

To many students there are only two problem solving skills: one is finding a suitable worked example to mimic, and the other is carrying out algorithmic computation. This is a consequence of our computationally based instruction with its emphasis on facts and techniques. To be an effective problem solver, one needs to consider alternatives, to experiment, to conjecture and test, and to analyze the results. An exploratory attitude, therefore, is a major problem solving skill. CASs encourage this type of attitude by providing graphical and numerical capabilities, as well as symbolic capabilities. Shifting the burden of computation to CASs makes time available for students to concentrate on how to approach a problem, to delineate subproblems, to consider alternatives, and to experiment.

Development of an exploratory approach

Students do not normally regard themselves as active participants in mathematical exploration, but rather as passive recipients of a fixed body of facts and algorithms. This is another consequence of the limitations imposed by hand computations. For example, students spend considerable time on convergence tests for series. However, having determined the convergence of a given series, the question of the sum (even in approximate form) is seldom considered. Likewise, questions about the rate of convergence are usually not raised. Our analysis is usually "one level deep" in the sense that we do not encourage students to pursue follow-up type questions. In fact, these types of questions are seldom even raised in our classrooms.

Computer algebra systems can be a major factor in developing an exploratory approach to learning. A characteristic of this approach is the generation of lots of examples (i.e., special cases). By making computations almost effortless, the student can be more easily encouraged to develop an experimental attitude. Computer aided analysis can become a standard process for investigating and developing conjectures.

Exercises

We can use CASs as tools to challenge our students with conceptual understanding, problem solving, and exploration. A few "CASs type" exercises are:

(a) Exploratory, open-ended questions such as developing heuristic methods for approximating an integral, summing a series, etc.

(b) Approximating within stated error bounds as well as determining error bounds for approximations (e.g., approximating the sum of a convergent series, or numerical integration).

(c) Making conjectures and then proving or disproving them (e.g., what are the geometrical interpretations of the multiplicities of the zeros, of numerator and denominator, of a rational function?). In a structured sequence of exercises, a student can be led to examine (with the aid of a CAS) enough cases to generate a conjecture.

(d) Construction of examples that satisfy certain constraints. Exercises in which students make up examples are an effective way to enhance conceptual understanding. CASs allow students to easily examine many cases and test out conjectures, helping students to either find the desired example or show why no such example can exist.

(e) Problems drawn from realistic models that are not feasible to do by hand computation.

(f) Problems invented by students whose answers can be checked by the CAS.

How to Use this Supplement

It is *not* expected that all sections will be covered, nor that the sections will be covered in a particular order. In some sections the techniques of calculus are applied to non-standard topics, such as fixed points, continued fractions, and recursive functions. The relative independence of the sections allows the instructor to tailor the material to fit the abilities of the class, the interests of the instructor, and the sequencing of the main calculus text.

Acknowledgments

We are grateful for family and friends who have been a constant source of support and encouragement for this project. We especially wish to recognize and thank Dave Kurtz, our colleague during the initial stages of development, for his inspiration and numerous contributions. Special thanks are extended to the following reviewers for their careful reading, thoughtful criticisms, and many valuable suggestions: Wade Ellis, Jr. (West Valley College), Arnold Ostebee (St. Olaf College), Phoebe Judson (Trinity University), Bonnie Gold (Wabash University), Herbert Brown (SUNY at Albany), John Devitt (University of Saskatchewan), John Ramsay (College of Wooster), Fred Schultz (Wellesley College).

Our appreciation is expressed to Richard Wallis and Robert Weinstein, McGraw-Hill Mathematics Editors, for their guidance, enthusiasm, and support. We would also like to acknowledge the support of the Alfred P. Sloan Foundation, Colby College, and the University of the South Pacific for our efforts.

Chapter 1

Introduction

1.1 What Are Computer Algebra Systems?

A Computer Algebra System (CAS) is a computer system for the exact solution of problems in symbolic form. This contrasts with the numerical analysis approach used in conventional computer languages such as FORTRAN or BASIC, where a numerical approximation is obtained. The ones of interest to us (Calculus T/L, DERIVE, MACSYMA, Maple, Mathematica, muMATH, and Theorist) are interactive systems which allow the user to define an expression, apply an operation, and manipulate the output. For example, if the user wishes to expand

$$(a + b)^{10}$$

the user applies the expansion operator:

    expand((a+b)∧10);

(where "∧" indicates exponentiation) obtaining as output the result, usually in two dimensional format.

The standard operations include the use of the system as an arbitrary precision desk calculator (do you want to know the factorial of 133?), algebraic simplification, calculus (differentiation, integration, power series), matrix algebra, systems of equations, differential equations, etc. There are also utility programs for expression manipulation, such as editing expressions and extracting parts of expressions. The systems may also allow numerical procedures such as numerical integration or graphing. If a procedure is not provided, most of these systems provide high-level programming capabilities to allow for user-written procedures as extensions to the system.

The initial CASs ran on mainframe computers and had a limited distribution. The most widely distributed system was REDUCE, which has been used in both Europe and the United States. Since the late 1970's several of the large systems (MACSYMA [9] and REDUCE) have been ported to smaller computers

1

and new large systems have become available for minicomputers (Maple [2] and Mathematica [7,?]) and personal computers (muMath [12] and DERIVE [5] on MS-DOS machines, and Calculus T/L and Theorist on Macintosh machines). Several large systems (Maple, Mathematica, Macsyma) are now available on personal computers with hard disks and 32 bit processors. Most CASs have a programming language that allows the user to define commands, although some of the smaller systems, such as DERIVE and Theorist, lack this capability.

Some references for general descriptions of CASs are [1,4,8,15]; for a discussion of the use of CASs in education, see [10,11]; for a comparison of different CASs, see [6].

1.2   Notation

The notation of the various CASs is similar to the standard mathematical notation. Each system allows the usual arithmetic operations and has procedures which perform many of the common mathematical functions. In describing Computer Algebra Systems, a writer can either use an actual CAS notation or develop a pseudo-notation. In this text we will primarily use the notation of the CAS Maple (version 4.2), since it is similar to many other notations, is reasonably complete, and is immediately understandable, by many readers. An appendix has some of the most commonly used Maple commands and the corresponding commands for other CASs. Many of the CASs have user's manuals or on-line help. The reader should see these sources for details.

The prompt from the CAS will be indicated by "> ". If a command is not finished when the user enters it, then the system will issue a secondary prompt, "≫ ", indicating that more is required from the user. The text entered by the user will be indented in "`typewriter`" font: Thus

> `factor(x∧2-1);`

is the command to factor $x^2 - 1$. Multiplication is indicated by $*$ and exponentiation by $\wedge$.

The set of all real numbers will be denoted by $\Re$. The set of all pairs of real numbers (which can be identified with the plane) will be denoted by $\Re^2$, and in general the set of all n-tuples of real numbers will be denoted by $\Re^n$.

We will put $\triangle$ at the end of an example, and $\square$ at the end of a proof. (We do have a few proofs!)

1.3   Advice on Using a CAS

On Using Different CASs

We have used the notation of the CAS Maple (version 4.2) in this text. We primarily use only the simplest commands, but if you have a different CAS (or a different version of Maple) you may need to modify the commands in the examples. The appendix contains tables to aid you in the translation process. To do a translation, find the Maple command we used in the Maple appendix. The corresponding command for the other CASs will be in the table for that CAS, with the same number and position in the table.

Example

Suppose you wish to define and plot the function $f(x) = \sin(x^2)$ over the interval $[-\pi, \pi]$ using the Mathematica CAS. In Maple, the commands are

```
> f:= proc(x) sin(x∧2) end;
> plot(f(x),x=-Pi..Pi);
```

To translate this into Mathematica, you look at the table in the Maple appendix and find the commands: **Define function** and **Plot**. Each table has two columns. The left-hand column describes the desired action, and the right-hand column gives the corresponding CAS command. They are listed in alphabetical order by the appropriate action. In this case Define function is command number 4. At the corresponding place in the Mathematica table you find the same action described in the left-hand column, and in the right-hand column In[n]:= **f[x_]:=** *expression*. As usual, the portion in **typewriter** font is entered by the user. The "In[n]:=" indicates the *n*th Mathematica prompt (which may be turned off in your Mathematica), and the **x_** indicates the variable.

Now we use the corresponding Mathematica commands (numbers 4 and 18) to define and plot our function.

In[]:= **f[x_]:= Sin[x∧2]**

From the table, the Mathematica plotting command has the form

Plot[*expression*, {*variable*, *first value*, *second value*}]

We enter the desired arguments:

In[]:= Plot[f [x], {x,-Pi,Pi}]

and obtain a plot similar to the Maple plot. △

Notational Conventions

The notation of the various computer algebra systems is similar, being based on mathematical notation. However, each has its own rules, especially with regard to the use of grouping symbols: parentheses "( )", brackets "[ ]", and curly braces "{ }". For example, Maple uses these symbols in the usual mathematical way: parentheses for function evaluation, e.g., $f(x)$, and for grouping; brackets for lists where order matters; and curly braces for lists where order doesn't matter, i.e., for sets. Mathematica uses brackets for function evaluation, e.g., $f[x]$, parentheses for grouping, and curly braces for lists, whether or not the order matters. We use the Maple notational conventions, since they are similar to those of mathematics. If you are using another system, you must take some care in understanding and using its conventions.

When Things Go Wrong

Computers (and their programs) are much like people, in that they sometimes make mistakes or respond in an unexpected way. However, unlike most people, they have unlimited patience: giving time to try experiments or waiting for an input while you think.

There are several things to keep in mind when using a CAS (or any computer program). It is very important to have an *exploratory attitude*. A CAS may not always do exactly what you expected, and several tries may be needed to obtain the desired result.

When doing calculations on decimal numbers, especially irrational numbers, only an approximate value is available. Each CAS will have an initial default number of digits of precision. In some calculations, the number of digits of precision may have to be increased. For example, in Maple, we set the number of digits of precision to 20 from the default 10 by

```
> Digits:= 20;
```

If a numerical calculation does not seem to give the correct results, try increasing the precision.

Graphing may also require some patience on the part of the CAS user. A plot may show too much of the curve, reducing the portion of interest to an insignificant smudge. Or, if the y-value goes to infinity, the range of y-values may need to be restricted. It is common practice to replot (adjusting horizontal and vertical scales) several times before obtaining a suitable picture.

The commands we use in this book are available on most computer algebra systems. However, your particular system may lack some command. It is also possible that some of the computations may require too much memory or time on your machine.

1.4   Getting Started

Prerequisites

- None

Discussion

The purpose of this section is to get you started in the use of a Computer Algebra System (CAS). The notation used in a CAS is similar to the standard mathematical notation and there are built-in functions which perform many of the common mathematical operations, such as the trigonometric functions, logarithms, etc. In this text we will primarily use the notation of the Maple CAS. The appendix *Computer Algebra Systems* has a brief introduction to several systems (including Calculus T/L, DERIVE, Maple, MACSYMA, Mathematica, muMATH and Theorist) with tables of corresponding commands for all systems except Calculus T/L and Theorist, which are visually oriented rather than command oriented. For additional information, many of the systems have on-line help as well as user manuals.

There are basically three ways in which a user interacts with a computer system or program: Interaction with a CAS *command line* (such as UNIX, VMS, and MS-DOS), *icons* (such as Macintosh), or *menu* (such as DERIVE). The menu systems will list some choices, and the user highlights the desired choice. Two Macintosh CAS programs, Calculus T/L and Theorist, are visually oriented: the user selects operations from a palette. In this manual we will assume a command-line CAS: they are the most common and are easier to describe in a text. You will have to refer to the locally available information on how to interact with your system.

Beginning and ending a session

To begin a session at the terminal, the user starts the program. This is done by typing in the name of the program or, in an icon system, clicking on the program's icon. When a program starts, a *prompt* will be displayed to let you know that it is ready for a command. In the examples in this text, the prompt symbol "> " will be displayed when a command is entered. We will use a semi-colon, ";", to indicate the end of a command. For example, to add 10 and 20 we enter

```
> 10+20;
```

(Note that the user input is in typewriter font.) The next thing you need to know is how to quit a CAS session. To do this, you enter a command (usually *quit*) or choose quit from the menu in an icon or menu system. (The appendix on the different computer algebra systems tells how to quit the various ones.)

Example 1

We will give an example of using a CAS, with all input and output shown. (After this section we will only show the interesting output.) You are encouraged to follow along on your CAS as you read – mathematics is not a spectator sport! We will use the syntax of Maple; see the appendix for translations of the commands into other CAS notations. In the next section we will discuss some of the operations in more detail.

This example will be a simple analysis of a polynomial,

$$x^4 - 6x^3 + 3x^2 + 26x - 24$$

After starting the system and getting a prompt we enter the polynomial and assign it a name, p. (Some CASs number their input and output for later reference. Others, like Maple, do not. In these CASs, if you wish to reuse output, then it is best to give it a name.)

```
> p:= x^4 - 6*x^3 + 3*x^2 + 26*x - 24;
```

$$p := x^4 - 6x^3 + 3x^2 + 26x - 24$$

The first line shows (in **typewriter** font) what the user entered. The second line shows the response by the CAS. The user enters the expressions in "1-dimensional" format, where exponentiation is indicated by a $\wedge$. The CAS responds in a 2-dimensional format. Now we *factor p*

```
> factor(p);
```

$$(x - 4)(x + 2)(x - 1)(x - 3)$$

and *expand* the result as a check:

```
> expand(");
```

$$x^4 - 6x^3 + 3x^2 + 26x - 24$$

The """ refers to the last expression displayed by the CAS. If you did not assign it a name, this allows you to recover it for further use. Having factored p, we have found its roots. As another, more general, way to find the roots of p: we solve the equation $p = 0$ for the variable x:

```
> solve(p=0,x);
```

$$4, -2, 1, 3$$

We now name the list of roots by "r":

```
> r:= ";
```

We can check this by substituting 4 in place of x in the expression p:

```
> subs(x=4,p);
```

$$0$$

It is easy to make a mistake and enter the parameters, "x=4" and "p", in the wrong order: `subs(p, x=4);`. This will result in an error message. The user should try an incorrect order to get some experience with error messages. Another common error is to use the wrong symbol at the end of the command. For example, if nothing is entered, the CAS Maple will wait until the semi-colon is entered; if a wrong symbol is entered, an error message will appear. In Maple, if a colon, "**:**", is entered instead of a semi-colon the work will be done, but nothing will be printed on the screen. This will result in a new prompt, but no output.

Finally, we will graph the polynomial on the interval $[-3, 5]$:

```
> plot(p,x=-3..5);
```

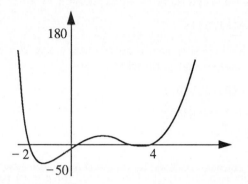

Graphics varies substantially among systems, so you should do some experimentation here: try graphing the polynomial with different domains (say x=-10..10) and with the various graphics options (e.g., restricting the y range or changing line styles). △

Operations

We will discuss some of the basic operations of Computer Algebra Systems in the rest of this section. Further operations are described as needed in the later sections. A list of all commands used (given for each CAS) is in the appendix. There are several other sources for information on commands: on-line help and the user's manual. You should get some experience using these sources. For example, most computer algebra systems will have built-in commands for the trigonometric functions and inverse trigonometric functions. The exact names used will vary with the system. For example, the inverse sine may be called *asin* or *arcsin*. To determine the name on your system, you can use the on-line help, the manual, or guess. To guess, just try entering the command using an argument with a known answer:

```
> asin(0);
```

which will return 0 if *asin* is defined. If it is not defined, then there will be an error message or the input *asin*(0) will be echoed (many computer algebra systems will simply copy the input back to the terminal when they cannot evaluate the expression). If *asin*(0) is not defined, enter *arcsin*(0).

```
> arcsin(0);
```

In using the trigonometric function it is useful to know that π is usually given a name (*Pi* in Maple), and that radians are used rather than degrees. Thus to find the sine of 60° or $\pi/3$ radians, one enters

```
> sin(Pi/3);
```

Sometimes the answer is exact, in terms of radicals, so $\sqrt{3}/2$ is returned. To obtain a decimal answer it is then necessary to evaluate the expression. In Maple, the command **evalf** evaluates an expression to a decimal number:

> `evalf(sin(Pi/3));`

will return .8660254040. If you need more accuracy, say 20 decimal digits, then you can use an optional second parameter:

> `evalf(sin(Pi/3),20);`

will return .86602540378443864675.

Exercises

1. Use the on-line help facility (or the manual, or guess) to find all the available trigonometric (and inverse trigonometric) functions. What are the names of the sine, arc sine, and other trigonometric functions? Construct a table of these.

2. With the functions found in the previous exercise, use the CAS to evaluate the following. Are the results what you expected? Explain.

   (a) $\arcsin(\sin(0))$

   (b) $\arcsin(\sin(\frac{\pi}{2}))$

   (c) $\arcsin(\sin(-\frac{\pi}{2}))$

   (d) $\arcsin(\sin(\pi))$

   (e) $\arcsin(\sin(-\pi))$

   (f) Try evaluating some of the other trigonometric functions.

Arithmetic

Every computer algebra system offers two types of numbers: exact integer and decimal. An integer can have arbitrarily many digits, limited by the storage capacity of the system. Computing the factorial of 5 is easy for human or computer:

> `5!;`

120

When first using a CAS it is important to experiment with problems with a known solution. This allows you to be sure that you are entering the data correctly and also allows you to check system behavior.

Now that we have confidence in our ability to enter the values in the correct format, and in the ability of the CAS to compute the correct answer, we can try a problem we would not attempt to do by hand. For example, a CAS can compute the factorial of 100:

```
> 100!;
```

> 9332621544394415268169923885626670049071596826438162\
> 4685929638952175999932299156089414639761565182862536\
> 79208827223758251185210916864000000000000000000000000\

The output is too long for one line, so it is continued on several lines, in this case with "\" indicating continuation.

The usual arithmetic operations are available: $+, -, *, /, \wedge, !$. The exponentiation operation is the circumflex, "$\wedge$", or double asterisks, "$**$": $2 \wedge 3 = 2 ** 3 = 2^3 = 8$. The division operation, "$/$", results in a rational number rather than a decimal. Rational arithmetic is also done exactly.

Decimal numbers can have high precision. Suppose you want to know the decimal expansion of π to 20 digits. You may first set the number of decimal digits to 20 by entering:

```
> Digits := 20;
```

and then evaluate π:

```
> evalf(Pi);
```

$$3.14159265358979323846$$

Until the digit length is reset, all future computations will now be given to 20 decimal digits. If greater accuracy is desired for just a single evaluation, it is best to use the optional second parameter in the **evalf** command, i.e., **evalf(Pi,20)**.

Exercises

3. What is 50!?

4. What is 60!?

5. How many digits are there in 40!, 50!?

Suppose we wish to continue investigating the number of digits in $n!$. Counting the number of digits is tedious and error prone. Perhaps there is a CAS command to help? Most CAS have on-line help; or you can use the manual. The user may have to guess the name of a command or topic.

6. Does your CAS have a command to determine the number of digits in an integer? What might be its name? [Some possibilities are "digit(s)," "size," "length," ....]

7. How does the number of digits in $n!$ grow? For example, if n is doubled, does the number of digits approximately double, triple, square, cube, or what?

8. What is the value of π, accurate to fifty digits?

Graphing

Most computer algebra systems have graphical as well as numerical and symbolic capabilities. It is often useful to graph a function to obtain information about it. Usually it is possible to graph a set of functions on the same axes by using a list. Such a graph is called a *multiplot*.

```
> plot({x^2,x},x = -1..1);
```

which returns the graph of the parabola $(y = x^2)$ and of the straight line $(y = x)$ over the interval $[-1, 1]$.

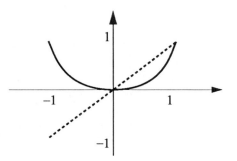

Now for an application of graphing. Suppose we wish to approximate the maximum of the function $f(x) = x^2 \sin(x) - x$ on the interval $[0, 3]$ with an error bound of 0.03. We define an algebraic expression, and give it a name for later reference:

```
> ex:= x^2*sin(x) - x;
```

(It is important to realize that this is an expression, and not a function: `ex(1)` does not make sense. See the next section on functions.)

One way to approximate the maximum over $[0, 3]$ is to graph it:

```
> plot(ex,x = 0..3);
```

We obtain the graph:

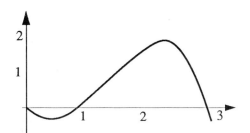

The maximum appears to occur between 2.0 and 2.5 with a value between 1.5 and 2.0. We can replot over a smaller interval to get a more accurate estimate:

```
> plot(ex,x = 2.0..2.5);
```

giving

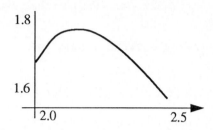

The maximum appears to occur between 2.1 and 2.2 with a value between 1.6 and 1.8. Continuing this process we can get as accurate an estimate as we want for the maximum.

Most CASs have a *point* option in their plot program that gives a point plot (in contrast to a line or curve plot). Point plots are very useful for suggesting the behavior of a sequence or of a function that is unbounded over the interval in question (e.g., rational functions). The Maple command for a point plot for the expression **ex** graphed above is

```
> plot(ex,2.0..2.5, style=POINT);
```

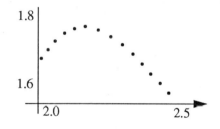

Exercises

9. Draw a point plot of the sequence $s(n) = \dfrac{15(n+1)\sin(n)}{n^2}$ and then, based on your plot, conjecture the value of $s(n)$ for large values of n.

10. Find the smallest positive root of the function $f(x) = x^2 \sin(x) - x$. (Clearly $f(0) = 0$. We want the smallest value greater than zero where f is zero.)

Factoring

If we are given a polynomial of degree n, i.e., $a_n x^n + a_{n-1} x^{n-1} + \cdots + a_1 x + a_0$, then we can attempt to find the roots by factoring. We are all familiar with the quadratic equation which gives the roots of a second degree polynomial. There are analogous formulas for the roots of third and fourth degree polynomials.

However, it is known that   no general formula exists for roots of polynomials
of degree greater than four. Most computer algebra systems have built-in procedures to attempt factoring, but they do not always succeed. Consider the
polynomial $x^5 - 17x^3 + 12x^2 + 52x - 48$. We give the polynomial a name and
try factoring it as before:

```
> poly:= x∧5 −17*x∧3 +12*x∧2 + 52*x −48;
> factor(poly,x);
```

$$(x - 3)(x - 2)(x - 1)(x + 2)(x + 4)$$

A CAS will return both real and complex roots if it can perform the factorization. If it cannot factor the polynomial, the CAS will print out the polynomial.
If factorization fails, then the graph can be sketched to approximate the roots,
as in the preceding exercise.

Exercises

11. Factor $2x^3 + 5x^2 + 4x + 1$.

12. Factor $6x^6 + 5x^5 + 5x^3 - 6x^2$.

13. Factor $6x^4 - \frac{7}{2}x^3 + \frac{5}{2}x^2 - \frac{7}{6}x + \frac{1}{6}$.

14. Factor $x^4 - 2x^3 + x^2 - x + \frac{6}{5}$.
     Hint: You may need to plot and "zoom in" (repeatedly) on portions of the
     x-axis in order to determine if there are any roots,

Solving Equations

We will show how to solve equations with the aid of a CAS, using an example
which can easily be done by hand. Again, when first using a CAS it is important
to experiment with problems with a known solution.

Suppose we wish to solve the following equation for x in terms of y:

$$\frac{4x^2 - y^2}{2xy + y^2} = 0$$

We begin by giving the expression on the left-hand side a name, say z:

```
> z:= (4*x∧2 − y∧2)/(2*x*y + y∧2);
```

We will use two approaches.

The first approach will mimic the method we might use without a computer.
We would notice that the equation is simplified if we multiplied both sides by the
denominator, leaving just the numerator. We can extract the numerator from a
ratio using the function **numer** and call the result *top*:

```
> top:= numer(z);
```

Now `top` is $4x^2 - y^2$. We can now solve the equation $top = 0$ for x in terms of y using the command `solve`:

> `solve(top = 0, x);`

$$y/2, \ -y/2$$

We should now examine our solutions to see if they make sense. If we substitute the first solution into `z`, we get what we expected:

> `subs(x=y/2,z);`

$$0$$

If we substitute the second solution into `z` we get:

> `subs(x = -y/2, z);`

$$\text{ERROR: } \tfrac{0}{0} \text{ undefined}$$

This means that z is of the form $\frac{0}{0}$ when $x = -y/2$. Expressions like $\frac{0}{0}$ are called *indeterminate forms*, since no numerical meaning can consistently be assigned. Thus the solution $x = -y/2$ is not a solution to our equation; it is an *extraneous solution*, one introduced by our method of solving the problem.

The second approach will be to allow the CAS to do all the work on its own, by applying `solve` to the original equation:

> `solve(z=0,x);`

$$y/2$$

We see that the CAS was smart enough to eliminate the extraneous solution $x = -y/2$. (Not every CAS may be so smart!) Is this solution valid for all values of y? No, we see that if we let $x = y = 0$ then the numerator and denominator are both zero:

> `subs({x = 0, y = 0}, numer(z));`

$$0$$

> `subs({x = 0, y = 0}, denom(z));`

$$0$$

(Note that we can make several substitutions at once.) If we substitute $x = 0, y = 0$ into the original expression we get the indeterminate form:

> `subs({x = 0, y = 0}, z);`

returns $\frac{0}{0}$. Thus even though the CAS returned the correct symbolic answer, the set of values for the variable y was not specified. *The user must interpret the answer to see when it makes sense.*

One way to eliminate extraneous solutions when the function is a rational function (a ratio of polynomials) is to factor the numerator and denominator first:

> `factor(z);`

$$\frac{2x - y}{y}$$

From this we can see that the solution to $\dfrac{4x^2 - y^2}{2xy + y^2} = 0$ is $y = 2x$ except when $x = y = 0$.

Systems of Equations

In many CASs the same command (`solve` here and in Maple) that can solve a single equation can also solve a system of equations. Usually the single equation is replaced by a list (or set) of equations and the single variable is replaced by a list (or set) of variables.

Suppose we want to solve the system of equations

$$x + 3y + 4z = 0$$
$$4x + 2y - 2z = 1$$
$$2x + y + z = 8$$

We enter the following command

> `solve({x+3*y+4*z=0, 4*x+2*y-2*z = 1, 2*x+y+z=8},{x,y,z});`

and obtain

$$\{z = \frac{15}{4}, y = -\frac{137}{20}, x = \frac{111}{20}\}$$

Well, you think, that was a linear system, and relatively easy to do by hand (if arithmetic errors are avoided). Let's try a non-linear system:

$$x^2 + 3x + y + 2z = 1$$
$$x + 4y^2 = 2$$
$$x - y + z = 1$$

by entering

> `solve({x^2+3*x+y+2*z=1, x+4*y^2=2, x-y+z=1},{x,y,z});`

which returns
$$\{z = 2, x = -2, y = -1\}$$

(as well as a solution in terms of the roots of polynomials it could not factor). Most non-linear systems cannot be solved exactly (instead, approximation techniques must be used), but sometimes a CAS can solve a non-linear system exactly.

Exercises

In Exercises 15 and 16 solve $f(x) = 4$ by doing the following:

(a) Remove common factors by factoring the rational function f

(b) Approximate a solution by graphing $y = f(x) - 4$ over smaller and smaller intervals

(c) Check your answer in (b) by using the solve command

15. $f(x) = \dfrac{x^3 - 3x^2/2 - x}{x^2 + x - 6}$

16. $f(x) = \dfrac{x^4 + 3x^3 + 2x^2}{x^2 + 3x + 2}$

17. $f(x) = \dfrac{x^3 + x^2/3 + 5x/9 + 1/9}{x^5 - x}$

18. Solve the following system:
$$x + 2y - 4z = 1, \qquad x + 4y - 2z = 2, \qquad x - y + z = 1$$

19. Solve the following system:
$$2x - 4y + 3z - 4w = 2, \qquad -x + 3y - 2z + w = 4$$
$$2x - y + z + 2w = 3, \qquad x + 2y - z + w = 1$$

20. Solve the following system (note that a CAS can find complex solutions):
$$x + y + 3z^2 = 1, \qquad x + y - z^2 = 3, \qquad 2x + 3y = 1$$

Functions

Computer Algebra Systems distinguish between *expressions* and *functions*. If we define

```
> ex:= 3*x + 1;
```

then **ex** is an expression, not a function. The statement **ex(1)** does not return the value 4. However, we can substitute values into expressions:

```
> subs(x=1,ex);
```

$$4$$

If we wish a function, then (in Maple) we would define the function $f(x) = 3x+1$ by:

```
> f:= proc(x) 3*x+1 end;
```

The command **proc** (as in "procedure") begins the function definition and **end** ends the function definition. Note that the variables are placed next to **proc** in the definition, not next to **f**. The appendix tells how to define functions using other Computer Algebra Systems. We use CAS functions in the usual way: **f(1)** is the number 4 obtained when evaluating **f** at 1:

```
> f(1);
```

$$4$$

Most CASs will give the value of a function in terms of radicals and simple functions, if possible. For example, if we define the function $f(x) = \sqrt{x}\,\sin(x+1)$

```
> f:= proc(x) sqrt(x)*sin(x+1) end;
```

and then evaluate at $x = 2$,

```
> f(2);
```

the response will be

$$2^{1/2}\sin(3)$$

If we wish a numerical value, we need to apply **evalf** to the output, **evalf(")**, or directly to the function:

```
> evalf(f(2));
```

returns .1995738293

Exercises

21. Define the function $f(x) = 3x^2 + 2x + 1$ using your CAS. Evaluate the function at several points to see if it is working correctly. Define an *expression* and see what the results are if you try using it as a function.

1.4.1   Control Structures

In our examples so far we have used a single command to obtain the desired result. Sometimes we need to perform a sequence of steps, with only minor changes between steps. A CAS may have built-in *control structures* to make such tasks easier. In this respect they are similar to standard computer languages like Pascal. Our discussion below will refer specifically to Maple, but most other Computer Algebra Systems have similar capabilities. The appendix gives the corresponding control structures in several CASs.

The for Loop

The **for** loop (similar to those in Pascal or FORTRAN) do a step a fixed number of times, with the loop control variable changing by a fixed amount (usually $+1$) on each step. The general form of a **for** loop in Maple is:

> **for** *control variable* **from** *starting value* **to** *final value* **do**
>>   *action to perform for each value of the control variable*
>> **od**;

The *action* is one or more commands, separated by semi-colons. The formatting doesn't matter: the entire command can be on one or several lines.

Example 2

Suppose we wish to see the cubes of the integers 5 through 10. Then we enter:

> **for** n **from** 5 **to** 10 **do**
>>   print(n∧3);
>> **od**;

The command **for** indicates the start of the loop and the command **od** indicates the end of the loop. The **for** loop control structure performs the step (printing n^3) with the control variable n equal to the values 5 through 10 (n being incremented by $+1$ after each step).   △

The while Loop

Sometimes the number of times a step is to be performed is not known in advance. A **while** loop performs a step while some condition is true. In a **while** command, we must increment any variables ourselves. The general form of the **while** loop in Maple is:

> **while** *condition* **do**
>>   *action while condition is true*
>> **od**;

The *action* is one or more commands, separated by semi-colons.

Example 3

Suppose we wish to find the smallest value of n with more than 10 digits in $n!$ (see the section on arithmetic above). We make a lower estimate of n, say 10; then while the length of $n!$ is less than or equal to 10 we print out the data and increment n by one:

> n:= 10;
> **while** length(n!) <= 10 **do**

```
≫          print(n, n!, length(n!));
≫          n:= n+1; #increment n by 1.
≫ od;
```

$$10, 3628500, 7$$
$$n := 11$$
$$11, 39916800, 8$$
$$n := 12$$
$$12, 479001600, 9$$
$$n := 13$$
$$13, 6227020800, 10$$
$$n := 14$$

Notice that the statement incrementing n is also printed out. To eliminate unwanted output, we can use a colon after **od** instead of a semicolon. This tells Maple to work silently, except when we explicitly tell it to print something. The **#** begins a *comment*: anything on the line after the **#** is ignored by Maple. Indentation is also ignored by Maple: we use it to set off the body of the loop step for clarity.   △

The if Structure

Sometimes during a loop we need to make a choice of one action or another. This is accomplished by means of an **if** statement. The general form of the statement is:

```
> if  condition then
≫          action if condition is true
≫ else
≫          action if condition is false
≫ fi;
```

The formatting does not matter: it can all be on one line.

Example 4

Suppose we wish to consider the piecewise defined function

$$f(x) = \begin{cases} 1 & \text{for } x < 2 \\ -x + 4 & \text{for } x >= 2 \end{cases}$$

We could enter:

```
> f:= proc(x) if x < 2 then
≫                  1 # Returns 1 as function value.
≫              else
≫                  -x+4 # Returns x as function value.
≫              fi
≫ end;
```

f can be evaluated at a particular point, i.e., $f(3) = 1$, but $f(x)$ has no meaning and thus cannot be displayed since our CAS does not know if x is less than 2 or greater than or equal to 2. Hence in the plot command the function is entered as f, not as $f(x)$. We now plot f over the interval $[0,5]$.

> `plot(f,x=0..5);`

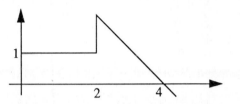

Surprise! Why does the graph contain the vertical line connecting the graphs of the two pieces of the function? The answer is that CASs plot by plotting points and connecting adjacent points with a straight line segment. Thus a CAS line or curve plot assumes the function is continuous. Try plotting the above function with a point plot. △

Piecewise defined functions with more than two pieces can also be defined and plotted.

Example 5

We define and then plot the function

$$f(x) = \begin{cases} -x^2 + 2 & \text{for } x < -1 \\ x^2 - 1 & \text{for } -1 \le x \le 1 \\ -(x-1)^2 & \text{for } 1 < x < 3 \\ x - 7 & \text{for } 3 \le x \end{cases}$$

We enter

```
> f:= proc(x) if x < -1 then −x² +1
≫              elif x<= 1 then x² − 1
≫              elif x< 3 then −(x − 1)²
≫              else x-7
≫              fi
≫ end;
```

We can now plot f.

> `plot(f,x=-3..5);`

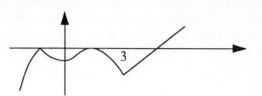

Again, the ability of plotting programs to handle non-differentiable functions will vary. △

Exercises

22. Use a **for** command to print the expansions of $(x + 1)^n$ for n between 2 and 8.

23. Use a **for** command to print $n!$ and length$(n!)$ for n from 1 to 50.

24. Use a **while** command to determine the smallest integer n such that length$(n!) \geq 2n$.

25. Use a **while** command to find the smallest integer n such that $n! > n^8 + 10$.

26. Graph the function
$$f(x) = \begin{cases} -x & \text{for } x < 0 \\ x^2 & \text{for } x \geq 0 \end{cases}$$

27. Consider the piecewise defined function
$$f(x) = \begin{cases} x^2 & \text{for } x < 0 \\ x & \text{for } 0 \leq x < 2 \\ -x + 4 & \text{for } 2 \leq x < 4 \\ x - 4 & \text{for } 4 \leq x \end{cases}$$

   (a) Plot the graph of f. Hint: Since f is a piecewise defined function, the command to plot f over the interval $[a, b]$ is: **plot(f, x=a..b)**. Note that the function is entered as f rather than as $f(x)$.

   (b) Define g to be the horizontal translation of f 2 units to the left, i.e., $g(x) = f(x + 2)$. (Enter **g:= proc(x) f(x+2) end;**) Draw a multiplot of f and g. (Enter **plot({f,g},x=-9..7);**)

Chapter 2

Functions

2.1   Elementary Graphing

Prerequisites

- None

Discussion

You have all had experience graphing individual functions, such as $f(x) = x^2$, $f(x) = \sin(x)$, and $f(x) = \log_{10}(x)$. With the aid of a CAS it is possible to examine many graphs easily, and to observe some patterns. Noting these patterns will allow you to recognize the graphs of several families of functions. And, given a member of a family, you can predict the nature of the graph. Often the family consists of functions which are transformations of the above familiar functions. The following exercises are designed to lead you to recognize patterns in the graphs of some functions. Using the CAS as a tool for graphing, you should experiment with other functions, as well as those given below. The exercises are grouped into several independent classes. The exercises within a class are generally independent, although the later exercises tend to build upon the earlier ones. Within each exercise, the parts should be worked in sequence. If you cannot answer a question, experiment with more functions.

Note that CASs recognize both $\log(x)$ and $\ln(x)$ as the natural logarithm. Recall that $\log_a(f(x)) = \dfrac{\log(f(x))}{\log(a)}$ where $a > 0$.

Exercises

Rational Powers

1. Integer powers: $f(x) = x^n$.

(a) For $x \in [-1, 1]$ graph $f(x) = x$, $f(x) = x^3$, and $f(x) = x^5$ on the same set of axes. Conjecture what the graph of $f(x) = x^{21}$ looks like. Check your conjecture with the CAS. In general, how does the graph of $f(x) = x^n$ vary with n, an odd positive integer? Does your conclusion change if the interval $[-1, 1]$ is increased?

(b) For $x \in [-1, 1]$ graph $f(x) = x^2$, $f(x) = x^4$, and $f(x) = x^6$ on the same set of axes. Conjecture what the graph of $f(x) = x^{20}$ looks like. Check your conjecture with the CAS. In general, how does the graph of $f(x) = x^n$ vary with n, an even positive integer? Does your conclusion change if the interval $[-1, 1]$ is increased?

(c) Perform the analysis of parts (a) and (b) with $f(x) = x^n$, n a negative integer. (Note: since $x^{-n} = 1/x^n$, you (or the CAS) should avoid the point $x = 0$.)

(d) Check your understanding of (a)-(c) by predicting the appearance of the graphs of $f(x) = x^7$, $f(x) = x^{-4}$, and $f(x) = x^8$. Check your answers with the CAS. Given the graph of $y = f(x)$, can you sketch the graph of $y = f(x)^n$ for any integer n?

2. Rational powers: $f(x) = x^r$.

(a) Sketch the graphs of $f(x) = x^{1/2}$ and $f(x) = x^{1/3}$ on the same axes for $x \in [-2, 2]$. How does the nature of the graph of $f(x) = x^{1/n}$ depend on n, a positive integer? Test your conjecture with $f(x) = x^{1/4}$ and $f(x) = x^{1/5}$.

(b) Sketch the graphs of $f(x) = x^{2/3}$ and $f(x) = x^{3/4}$ on the same axes for $x \in [-2, 2]$. How does the nature of the graph of $f(x) = x^{m/n}$ depend on positive integers m and n? Test your conjecture with $f(x) = x^{4/7}$ and $f(x) = x^{3/8}$.

Transformations of Graphs

3. Reflections of graphs.

(a) Sketch the graphs of $f(x) = x^2$, $f(x) = -x^2$, $f(x) = x^3$, and $f(x) = -x^3$ on the interval $[-2, 2]$. How does the graph of $y = -f(x)$ depend geometrically upon the graph of $y = f(x)$? Can you describe how to get from one to the other by a reflection? Test your conclusions on $f(x) = \sin(x)$ and $f(x) = \log_{10}(x)$.

(b) Sketch the graphs of $f(x) = x^2$, $f(x) = (-x)^2$, $f(x) = x^3$, and $f(x) = (-x)^3$, on the interval $[-2, 2]$. How does the graph of $y = f(-x)$ depend geometrically upon the graph of $y = f(x)$? Can you describe how to get from one to the other by a reflection? Test your conclusions on $f(x) = \sin(x)$ and $f(x) = (x + 1)^2$.

4. Horizontal and vertical translations.

(a) Sketch the graphs of $f(x) = x^2$, $f(x) = (x-1)^2$, $f(x) = (x-2)^2$, $f(x) = (x+2)^2$, and $f(x) = (x+1)^3$ on the interval $[-3,3]$. How does the graph of $y = f(x+a)$ depend upon the graph of $y = f(x)$ and a? Test your conjecture with $f(x) = x^{1/2}$ and several positive and negative values of a.

(b) Sketch the graphs of $f(x) = x^2$, $f(x) = x^2 + 1$, $f(x) = x^2 - 2$, and $f(x) = x^3 + 1$ on the interval $[-3,3]$. How does the graph of $y = f(x) + a$ depend upon the graph of $y = f(x)$ and a? Test your conjecture with $f(x) = x^{1/2}$ and several positive and negative values of a.

(c) Combine your results from (a) and (b) to make a statement on how the graph of $y = f(x+a) + b$ depends upon the graph of $y = f(x)$, a and b. Test your conclusion with $f(x) = \log_{10}(x)$, $a = 1$, $b = 2$.

5. Scaling (stretching) graphs.

(a) Sketch the graphs of $f(x) = x^2$, $f(x) = 2x^2$, $f(x) = (x-1)^2$, $f(x) = 2(x-1)^2$, $f(x) = \log_{10}(x)$, and $f(x) = 2\log_{10}(x)$ on $[-2,2]$. How does the graph of $y = af(x)$ depend upon the graph of $y = f(x)$ and a. How does the sign of a affect the result? Test your answer to this question on $f(x) = (x-1)^{1/2}$, $a = -2$, and $f(x) = (x+1)^2$, $a = \frac{1}{2}$.

(b) Sketch the graphs of $f(x) = x^2$, $f(x) = (2x)^2$, $f(x) = (\frac{1}{2}x)^2$, $f(x) = (x-1)^2$, and $f(x) = (2x-1)^2$ on $[-2,2]$. How does the graph of $y = f(ax)$ depend upon the graph of $y = f(x)$ and $a > 0$? How do the sign and magnitude of a affect the result? Test your answer to this question on $f(x) = (x-1)^{1/2}$, $a = -2$.

6. Based on your results in Exercises 3 through 5, how do the graphs of $y = f(x) + a$, $y = f(x+a)$, $y = af(x)$, and $y = f(ax)$ depend upon the graph of $y = f(x)$ and a. How does the sign of a affect the result?

7. Investigate combinations of the above transformations. Given the graph of $y = f(x)$, the graph of $y = af(bx+bc) + d$ is obtained by applying the transformations: (1) scale horizontally by replacing x by bx, giving $y = f(bx)$, (2) shift horizontally by replacing the independent variable x by $x+c$, giving $y = f(bx+bc)$, (3) scale vertically giving $y = af(bx+bc)$, and finally (4) shift vertically giving $y = af(bx+bc) + d$. Does the result depend upon the order in which these operations are carried out? Test with some of the standard functions.

8. What sequence of transformations (see Exercise 7) can be used to obtain the graph of $y = 3(x+1)^2 + 4$ from the graph of $y = x^2$? What sequence of transformations can be used to obtain the graph of $y = 2(3x+6)^2 + 5$ from the graph of $y = x^2$? Test your sequence using the substitution and graphing commands of the CAS.

9. Applications to trigonometric functions.

   (a) As an application of Exercise 4, what do the graphs of $f(x) = \sin(x + \pi)$ and $f(x) = \sin(x + \frac{\pi}{2})$ look like?

   (b) As an application of Exercise 5, what do the graphs of $f(x) = \sin(3x)$ and $f(x) = 3\sin(x)$ look like?

   (c) As an application of Exercise 7, what does the graph of $f(x) = 5\sin(2x + 4)$ look like? (The function $f(x) = A\sin(Bx + BC)$ describes a wave with *amplitude* A, *period* $2\pi/B$, and *phase* C.)

   (d) Sketch the graph of $f(x) = \sin(x) + \cos(x)$. Does this appear to be of the form $f(x) = a\sin(x + b)$ for some a and b? Try to find suitable values of a and b by sketching graphs and applying trigonometric identities.

Exponential and Logarithmic Functions

10. General appearance of exponential functions.

   (a) For $x \in [-2, 2]$ sketch the graphs of $f(x) = 2^x$, $f(x) = 3^x$, $f(x) = (0.5)^x = 1/2^x = 2^{-x}$, $f(x) = 10^x$, and $f(x) = 10^{-x}$. Which of these are increasing and which of these are decreasing?

   (b) In general, what is the appearance of the graph of $f(x) = a^x$ for the cases $0 < a < 1$, $1 < a < \infty$, $a = 1$, and $a < 0$? Write down a general statement covering these cases and test it using examples.

11. Comparison between exponential and power functions.

   How do you think that functions of the form $f(x) = x^n$ and $f(x) = a^x$ compare for large values of x? If $n > 1$ and $a > 1$, then they are both increasing. Which increases at a faster rate? Experiment with some functions. Does $f(x) = 2^x$ eventually become larger than $g(x) = x^{100}$, and remain larger? You can use the CAS as a calculator to evaluate both functions (or the difference, $h(x) = f(x) - g(x)$). Test your conclusions with other values of a and n.

12. General appearance of logarithmic functions.

   (a) For $x \in [0.1, 10]$ sketch the graphs of $f(x) = \log_{0.5}(x)$, $f(x) = \log_{0.1}(x)$, $f(x) = \log_2(x)$, and $f(x) = \log_{10}(x)$. Now try these functions for the intervals $[0.01, 10]$ and $[0.001, 10]$. What happens to $\log_a(x)$ as x gets close to zero?

   (b) What is the relation between the graphs of $f(x) = \log_a(x)$ and $f(x) = \log_{1/a}(x)$?

   (c) For what values of a is $f(x) = \log_a(x)$ an increasing function? Test your conclusion by graphing some examples.

13. Comparison between logarithmic and power functions.

For $x \in [1, 10]$ sketch the graphs of $f(x) = \log_{10}(x)$ and $g(x) = x^{1/2}$. They are both increasing, but levelling out. What happens as x gets large? Graph these functions for large x, make a guess as to the general result, and test your guess with some more examples of the form $f(x) = \log_a(x)$, $a > 1$ and $g(x) = x^r$, $0 < r < 1$.

2.2 Data Fitting

Prerequisites

- An elementary knowledge of functions and graphing

Discussion

An analysis of an on-going process usually involves sampling the process. The sample values (called *data points*) consist of ordered pairs of numbers. The first number in each pair is the value of the control variable, and the second number is the value of the process associated with the value of the control variable. For example, hourly temperature readings or closing stock market averages or voter preference polls, etc., are examples of data sets where the control variable is time. Data fitting is the process of developing a mathematical description (i.e., a model) from a set of data values that is suitable for predictive purposes. How to do this is an important and often difficult question. There are several methods that can be used, each of which gives an approximate description. The choice of method is usually dependent on the nature of the process being described and the accuracy desired.

We will consider a process that can be approximately described by a polynomial. We assume that there is an (unknown) polynomial such that each process value is obtained by evaluating the polynomial at a control variable value. Thus each control variable value in the data set has exactly one corresponding process value. (This reflects the fact that a polynomial cannot have two different values at one point.) Plotting the data set in the plane, our task becomes one of finding a polynomial whose graph "fits" the set of points. We shall restrict ourselves to methods of data fitting that require the graph of the polynomial to contain the data points. (Thus we will not, for example, consider the method of least squares.)

Constructing the desired polynomial consists of determining the values for the polynomial's coefficients as described and illustrated in the following examples.

Polynomials

A general second degree polynomial has the form $ax^2 + bx + c$. The unknown coefficients a, b, and c are called the *parameters* of the polynomial. A third degree polynomial has the form $ax^3 + bx^2 + cx + d$ with parameters a, b, c, and d.

Example 1

We will fit a second degree polynomial to the set of data points $\{(1,5), (3,1), (6,6)\}$ by determining the parameters in the general second degree polynomial $ax^2 + bx + c$. Since we want the graph of the polynomial to pass through each of the three data points, we begin by defining a system of three equations in the function $p(x) = ax^2 + bx + c$ whose unknowns are the parameters a, b, and c in the following manner:

$$p(1) = 5 \quad p(3) = 1 \quad p(6) = 6$$

The solution of this set of three equations in three unknowns gives the values of parameters a, b, and c.

The commands for solving this problem using the Maple CAS are:

```
> p:= proc(x) a*x^2 + b*x + c end;
> solve({p(1)=5,p(3)=1,p(6)=6},{a,b,c});
> q := subs(", p(x));
```

The result is $q = \frac{11}{15}x^2 - \frac{74}{15}x + \frac{46}{5}$. (Note that q is an expression, not a function, since it is obtained by substituting values for the parameters a, b, and c in the expression $p(x)$. Thus an attempt to evaluate q will not give the desired results. Most CAS commands return expressions rather than functions.)   △

We observe that when the number of parameters is equal to the number of data points, a unique polynomial is determined that has degree one less than the number of data points. (Why is this true?)

What happens when there are more parameters than data points? For example, how can you fit a third degree polynomial to a set of three data points? The answer is to arbitrarily assign a value to one of the parameters and then use the three data points to determine the values of the remaining three parameters. In general, we fit an nth degree polynomial (with $n+1$ parameters) to a set of k data points ($k \le n+1$) by assigning arbitrary values to $n+1-k$ of the parameters and then use the k data points to determine the values of the remaining k parameters. The only restriction is that the zero value cannot be assigned to the parameter that is the coefficient in the highest power term, since this would reduce the degree of the polynomial.

When there are fewer parameters than data points, a number of data points equal to the number of parameters is selected. Thus the process of fitting a polynomial to a set of data points gives a unique result only in the case where the number of parameters is equal to the number of data points.

Example 2

We will fit a third degree (cubic) polynomial to the set of data points in Example 1.

Since the general third degree polynomial $(ax^3 + bx^2 + cx + d)$ has four parameters and we have only three data points, we must assign an arbitrary value to one of the parameters b, c, or d. We will let $c = 0$. We now proceed as in Example 1. That is, we define our function to be $p(x) = ax^3 + bx^2 + d$, determine a system of equations by evaluating our polynomial function at each of the data points, solve the system of equations for the parameter values, and substitute these values into the definition of $p(x)$, obtaining as our result an expression that we label q. The Maple CAS commands are:

```
> p:= proc(x) a*x^3 + b*x^2 + d end;
> solve({p(1)=5,p(3)=1,p(6)=6},{a,b,d});
> q := subs(", p(x));
```

The result is $q = \frac{74}{405}x^3 - \frac{443}{405}x^2 + \frac{266}{45}$. △

Note that for a unique solution:

a. The number of data points must be the same as the number of undefined coefficients in the polynomial. Thus the number of data points is one more than the degree of the polynomial.

b. The evaluation of the polynomial (with unknown coefficients) at a *single* data point gives a linear equation whose variables are the unknown coefficients.

c. The values of the coefficients are the solutions of the *system* of equations determined by evaluating the polynomial at each of the data points.

d. The graph of the polynomial obtained by substituting the solutions for the unknown coefficients passes through each of the data points.

Exercises

1. Fit a third degree polynomial to the set of points $\{(-2,7), (1,1), (5,9), (7,4)\}$.

2. Assume that there is a linear relation between Celsius and Fahrenheit temperatures. Determine a linear function (i.e., a first degree polynomial) that expresses Fahrenheit temperature as a function of Celsius temperature. Hint: Convert this problem into a curve fitting problem by determining two data points of the form (Celsius temperature, Fahrenheit temperature). For example, consider the respective freezing temperatures (0,32) and the respective boiling temperatures (100,212).

3. Using data values from the following height (in) versus weight (lb) table,

Height	Weight	Height	Weight
60	132	71	185
61	136	72	190
62	141	73	195
63	145	74	201
64	150	75	206
65	155	76	212
66	160	77	218
67	165	78	223
68	170	79	229
69	175	80	234
70	180		

(a) Fit a first degree polynomial to the set of data points $\{(60, 132), (80, 234)\}$. Evaluate your polynomial at $x = 63, 68, 72$, and 78 and compare the results against the weights listed in the table.

(b) Select five values from the table and fit a fourth degree polynomial to the corresponding set of data points. Evaluate your polynomial at two values of x (not used in determining the polynomial) and compare the results against the weights listed in the table.

(c) Draw a multiplot of the polynomials found in parts (a) and (b), i.e., superimpose the graphs of the two polynomials on the same set of axes.

4. Consider the sine function, $f(x) = \sin(x)$, over the interval $[0, \pi]$.

(a) Approximate $f(x) = \sin(x)$ with a second degree polynomial. Hint: Convert this problem into a data fitting problem. Determine three data values by evaluating $f(x) = \sin(x)$ at three different points in the interval $[0, \pi]$, say $x = 0, \frac{\pi}{2}, \pi$.

(b) Determine the largest error that can occur in approximating $f(x) = \sin(x)$ with your second degree polynomial. Hint: Draw a multiplot of $f(x) = \sin(x)$ and your second degree polynomial and determine by inspection the largest *vertical* distance between the two curves.

(c) Repeat parts (a) and (b) choosing a different middle point.

(d) What is the size of the error when f is approximated by the third degree polynomial $x^3 + bx^2 + cx + d$ fitted to the three data points used in part (a)?

5. Suppose that yesterday you bought a snack at the Spa at one and three in the afternoon and at seven and ten at night. You observed that there were three people in line ahead of you at one o'clock, none at three o'clock, six at seven o'clock, and two at ten o'clock. Can you approximate from these observations how many people would have been ahead of you in line if you

had gone to the Spa at four o'clock? eight o'clock? Hint: Convert the problem into a data fitting problem, fit a third degree polynomial to your data, and then evaluate the third degree polynomial at $x = 4$ and $x = 8$.

6. This is an exploratory type exercise. Consider the cosine function $f(x) = \cos(x)$ over the interval $[-\frac{\pi}{2}, \frac{\pi}{2}]$ and the three data values obtained by evaluating f at $x = -\frac{\pi}{2}, 0, \frac{\pi}{2}$. Experiment with approximating f by a cubic polynomial $ax^3 + bx^2 + cx + d$ fitted to the three data points when

   (a) The coefficient a is assigned the value 1. Draw a multiplot of f and the cubic polynomial. How would the plot of the cubic polynomial change if the coefficient a were given a larger value? a negative value?

   (b) The coefficient c is assigned the value 1. Draw a multiplot of f and the cubic polynomial. How would the plot of the cubic polynomial change if the coefficient c were given a larger value? a negative value?

   (c) The coefficient b is assigned the value 1. Explain the result.

7. This is an exploratory type exercise.

   (a) Fit polynomial functions of degrees three, four, five, and six to the waiting-in-line data from Exercise 5.

   (b) Draw a multiplot of your four functions (i.e., superimpose the graphs of the four functions on the same set of axes).

   (c) Which of the four polynomials would give the best approximation of the number of people who would have been in line ahead of you at four o'clock? eight o'clock? Explain your reasoning.

   (d) Make a conjecture concerning the degree of the data fitting polynomial and the accuracy of the polynomial in predicting results at intermediate points between the data points. Test your conjecture by applying it to one or two other "real life" situations.

2.3 Polar Coordinates

Prerequisites

- Ability to graph functions in polar coordinates

Discussion

The use of polar coordinates allows for the analytic description and analysis of families of curves that would be difficult to handle using rectangular coordinates. The object of this section is to examine and discover some properties of functions described using polar coordinates.

If a curve is described as the graph of a function in rectangular coordinates, then the graph cannot have any closed loops (such as a circle), since for a given x

value there can be at most one corresponding y value. However, using polar coordinates, curves with loops can be described as graphs of functions. For example, $r = 2$ describes a circle of radius 2 centered at the origin, and $r = f(\theta) = \sin(\theta)$ describes a circle with radius 1 centered at $(\frac{1}{2}, 0)$.

Example 1

We will consider the family of curves given by

$$r = a\theta + b$$

We will examine a sequence of examples, looking for a pattern. To do the graphing, we use our CAS. Most CASs will have a polar coordinate graphing option. In Maple to plot $r = a\theta + b$ with $0 \leq \theta \leq 6$ we enter:

```
> plot([a*t+b,t,0..6],coords=polar)
```

where t is used in place of θ (not commonly found on keyboards).

Now let's examine a few plots. First we set $b = 0$ and vary a. When $a = 1$ we obtain the graph of $r = \theta$ on the left; when $a = 2$ we obtain the graph of $r = 2\theta$ on the right.

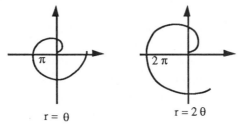

We see that the general shape is a spiral, and that multiplying θ by 2 increases the rate at which the spiral expands (note that the rate of increase $dr/d\theta = 2$).

Next we keep a constant and vary b. If we let $b = 1$ we get the graph on the left; when we let $b = -1$ we get the graph on the right:

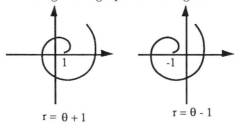

Thus adding a constant b shifts the origin to the point b on the x-axis.

Notice that all of the spirals open in a counter-clockwise direction. What if we want our spiral to open in a clockwise direction, as on the left below:

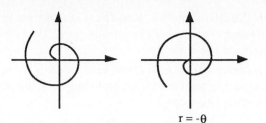

$$r = -\theta$$

If we try $r = -\theta$ we get the plot on the right; not what was wanted – it still opens counter-clockwise. What should we do? Experiment! (The solution is left to the reader.)   △

Exercises

In the following exercises, you are asked to discover patterns in the graphs.

1. The graph of $r = f(\theta) = \sin(n\theta)$.

   (a) Sketch the graph of $r = \sin(\theta)$.
   (b) Sketch the graph of $r = \sin(2\theta)$.
   (c) Sketch the graph of $r = \sin(3\theta)$.
   (d) Sketch the graph of $r = \sin(4\theta)$.
   (e) What can you conclude about the graph of $r = f(\theta) = \sin(n\theta)$ for n positive?
   (f) What can you say about the graph of $r = f(\theta) = \sin(n\theta)$ for negative n?

2. Formulate and test a conjecture about the graph of $r = f(\theta) = \cos(n\theta)$, n both positive and negative. You may wish to graph the function for small values of n as in Exercise 1.

3. Formulate and test a conjecture about the graph of $r = f(\theta) = \tan(n\theta)$, n both positive and negative. You may wish to graph the function for small values of n as in Exercise 1.

4. Formulate and test a conjecture about the graph of $r = f(\theta) = a + b\sin(\theta)$. What roles do the parameters a and b play? As in the previous exercises, you may wish to try graphing the function with several values of a and b.

5. Formulate and test a conjecture about the graph of $r = f(\theta) = a + b\cos(\theta)$. What roles do the parameters a and b play?

6. The graph of $r = f(\theta) = k \cos(\theta) - \cos(n\theta)$.

   (a) Sketch the graph of $r = k\cos(\theta) - \cos(\theta)$ for several values, both positive and negative, of the parameter k. What effect does the value of k have on the curve?

(b) Sketch the graph of $r = k\cos(\theta) - \cos(2\theta)$ for several values, both positive and negative, of the parameter k. What effect does the value of k have on the curve?

(c) Sketch the graph of $r = k\cos(\theta) - \cos(3\theta)$ for several values, both positive and negative, of the parameter k. What effect does the value of k have on the curve?

(d) What effect do the values of k and n have on the curve $r = k\cos(\theta) - \cos(n\theta)$?

7. Repeat Exercise 5 for the curve $r = k\sin(\theta) - \cos(n\theta)$.

In Exercises 8-11, remember that the transformation equations from rectangular to polar coordinates are:

$$x = r\cos(\theta), \quad y = r\sin(\theta), \quad r = \sqrt{x^2 + y^2}$$

8. Let L be the line in the plane $x = -1$; let r be the distance of a point P to the origin; let d be the distance of P from L.

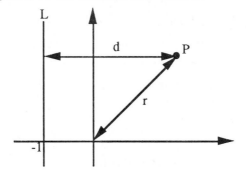

Find the equation in polar coordinates of all points P such that $r + d$ is a constant k, positive, negative, or zero. What effect does k have on the shape of the curve?

9. With the same situation as in Exercise 8, find the polar equation of the curve of all points P such that the difference $r - d$ is a constant k.

10. With the same situation as in Exercise 8, find the polar equation of the curve of all points P such that the product rd is a constant k.

11. With the same situation as in Exercise 8, find the polar equation of the curve of all points P such that the ratio r/d is a constant k.

2.4   Roots (Zeros) of Polynomials

Prerequisites

• Polynomials in one variable

- Section on Elementary Graphing

Discussion: Part 1

This section investigates a few properties of polynomial functions $p : \Re \longrightarrow \Re$. Such a p has the form

$$p(x) = a_n x^n + a_{n-1} x^{n-1} + \cdots + a_1 x^1 + a_0$$

where the $a_0, \ldots, a_n$ are real numbers. The (non-zero) integer n is the *degree* of the polynomial. A *zero (root)* of a function f is a value x such that $f(x) = 0$. When the function in question is a polynomial, the term *root* rather than zero is usually used. We will investigate roots (zeros) of polynomials. The first major result is the *Fundamental Theorem of Algebra* which says that a polynomial of degree n has n roots. Not all of these roots need be distinct real numbers. Some of the roots may be complex numbers: for example, the polynomial

$$p(x) = x^2 + 1$$

has two complex roots, $i = \sqrt{-1}$ and $-i = -\sqrt{-1}$. Some of the real roots may not be distinct: for example, the polynomial

$$p(x) = x^2$$

has two identical real roots, $x = 0$ and $x = 0$. The root 0 is said to be of *multiplicity* 2. The polynomial

$$p(x) = x^2 (x - 1)^3$$

is of degree 5. Zero is a root of multiplicity 2 and 1 is a root of multiplicity 3. In general, if a root occurs exactly m times in a factorization, then it is a root of multiplicity m. Thus if the polynomial p of degree n has k distinct roots, $r_1, ..., r_k$, then p can be factored

$$p(x) = (x - r_1)^{n_1} \cdots (x - r_k)^{n_k}$$

Here $n_1 + \cdots + n_k = n$ and the multiplicity of the root r_i is n_i.

Exercises

1. Give an example of a polynomial with roots 1, 2, and -3 of multiplicities 3, 2, and 4 respectively.

2. Give examples of five polynomials that have 2 as a root of multiplicities 1, 2, 3, 4, and 5 respectively.

3. Sketch the graphs of the polynomials in Exercise 2.

4. Frame a conjecture that relates the multiplicity of the root to the nature of the graph near the root. That is, what simple regularities do you observe in the graph as the multiplicity changes from $1, 2, \ldots$?

5. Consider the situation when the polynomial is the denominator of a rational function, e.g., $r(x) = \frac{1}{p(x)}$. What is the geometrical interpretation of a multiple root of p in this situation?

Discussion: Part 2

We will begin considering some methods of finding roots of polynomials in this section. The topic is also treated in later sections.

The first approach will be to factor the polynomial. If the degree of the polynomial is less than 5, then there are formulas (such as the quadratic formula) which can be applied. There are no formulas for polynomials of degree 5 or higher, but a CAS can handle some of these.

Example 1

If we wish to factor

$$x^5 - 8x^4 - 55x^3 + 410x^2 + 744x - 4032$$

we enter

```
> factor(x∧5 - 8*x∧4 - 55*x∧3 + 410*x∧2 + 744*x - 4032);
```

and get the response

$$(x - 3)(x + 4)(x - 8)(x + 6)(x - 7)$$

This was easy for the CAS since the roots were small integers. If we try factoring

$$x^5 + 3x^4 + 2x + 10$$

the CAS will probably fail. What can we do?   △

Our second approach will be to graph the polynomial. We can approximate a root with as much accuracy as we like if we can graph the polynomial over a small enough interval containing the root. The complex roots of a polynomial p must come in conjugate pairs, so the number of complex roots must be even.

Example 2

We will try graphing the polynomial $p = x^5 + 3x^4 + 2x + 10$, for which factoring failed. Since the degree of p is odd, there must be at least one real root. (Why?) For large values of x, p is dominated by the term of highest degree, i.e., x^5 which is of odd degree. Thus when x is very positive, $p(x)$ is a large positive number. Similarly, when x is very negative so is $p(x)$. For a continuous function to go from negative to positive it must pass through zero (Intermediate Value Theorem). Thus a polynomial of odd degree must have at least one real root. We graph p on $[-5, 5]$:

```
> p:= x^5 + 3*x^4 + 2*x + 10;
> plot(p,x=-5..5);
```

This gives us the graph in (a):

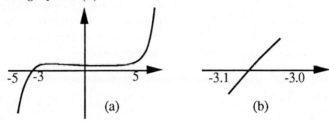

(a)   (b)

We see that p has a real root, near -3. Graphing p over the smaller interval $[-3.1, -3.0]$ gives us the graph in (b). We can continue this process until we have the desired accuracy for the root (or until the plotter's precision is exceeded). Is this the *only* real root? It is clear that $p(x)$ is positive for positive values of x. (Why?) It is also clear that $p(x)$ is negative for $x < -5$. (Note that for $x < -5$ the sum of the first two terms is negative and the sum of the last two terms is negative.) Thus the root near -3 is the *only* real root of p. Could it be a root of multiplicity 3? multiplicity 5? How can you tell? If the root had multiplicity 3 or 5, would p have to have an x^2 term? (Why?)

$\triangle$

Example 3

We want to find all real roots of $p = x^5 - 8x^4 + x + 11$. Again, p cannot be factored. Since the degree of p is 5, there must be 1, 3, or 5 real roots. (Why?) We graph p on some interval, say $[-10, 10]$ obtaining (a) below:

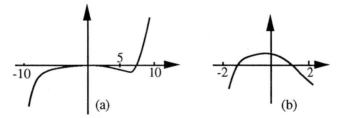

(a)   (b)

We can see that there is a root between 5 and 10. From the graph it is difficult to tell what is happening near zero, but from the shape of the curve and since $p(0) = 11 > 0$, there must be at least 2 distinct real roots near zero. To get a clearer picture of the shape of the curve near zero, we plot p over a smaller interval, say $[-2, 2]$. The graph is shown in (b). It is clear that there is a root in the interval $[-2, 0]$ and another in $[0, 2]$. We can obtain better approximations

of the three real roots as before. In the section on the Bisection Algorithm we
discuss a method to obtain very accurate approximations of zeros once they have
been isolated by graphing.   △

The procedure for graphically approximating the roots of a polynomial is to
first determine the possible number of real roots and then to plot the polynomial
over an interval, say the default interval. Frequently it is necessary to replot
(zoom) three or four times in order to isolate the roots. Once isolated, a root
can be approximated by repeated replotting (zooming) as illustrated in Example
2 or by employing a root finding algorithm such as the Bisection Algorithm. It
is important to consider the multiplicities of the isolated roots as well as the
possibility of roots that lie outside the original interval.

Exercises

In Exercises 6 through 8, find all real roots of the polynomial with an accuracy
of at least 0.01.

6. $x^5 - 3x^3 + 2x + 1$

7. $x^6 + x^5 - x^2 - 5$

8. $x^6 + 6x^5 - 16x^4 - x^2 - 6x + 17$

In Exercises 9 through 11 the process will be reversed: you will construct polynomials with the desired property.

9. Construct a polynomial with real roots at -2, -1, 3, and 4.

10. Construct a polynomial with two complex roots and with real roots at -2
    and 1.

11. Construct a polynomial with integer coefficients that has exactly two com-
    plex roots and exactly five distinct real roots, none of which are integers.

12. Construct a non-negative polynomial (i.e., a polynomial that has no neg-
    ative values) that has roots at -1 and 3.

2.5   Even and Odd Functions

Prerequisites

- Some experience in graphing functions

Discussion

A function $f : \Re \longrightarrow \Re$ is *even* if $f(x) = f(-x)$ for all x in its domain. An *odd* function satisfies $f(x) = -f(-x)$ for all x in its domain. As examples:

$$f(x) = 1 \qquad \text{even}$$
$$f(x) = x \qquad \text{odd}$$
$$f(x) = \sin(x) \qquad \text{odd}$$
$$f(x) = \cos(x) \qquad \text{even}$$
$$f(x) = 1/x \qquad \text{odd}$$
$$f(x) = 0 \qquad \text{even and odd}$$
$$f(x) = x + 1 \qquad \text{neither}$$

We will illustrate two ways in which a CAS can be used to determine if a function is even, odd, or neither by verifying that the function $f(x) = \cos(x)$ is even. The first method will use the symbolic aspect of a CAS and the second method the graphic aspect of a CAS.

We begin by defining $f(x) = \cos(x)$.

```
> f:= proc(x) cos(x) end;
```

We now apply the definition in the form: f is an even function provided $f(x) - f(-x) = 0$.

```
> simplify(f(x) - f(-x));
```

This gives the result 0, verifying that f is an even function.

Now consider the graph of an even function. Since $f(x) = f(-x)$ the graph must be symmetric about the y-axis. That is, the portion of the graph lying to the left of the y-axis is the reflection in the y-axis of the portion of the graph that lies to the right of the y-axis. To check for this symmetry, we use a CAS to obtain a multiplot of $f(x)$ and $f(-x)$, i.e., plot the graphs of $y = \cos(x)$ and $y = \cos(-x)$ on the same set of axes over the interval $[-8, 8]$. The CAS command is

```
> plot({f(x), f(-x)},x=-8..8);
```

Note how the cursor drawing the second graph moves along the curve of the first graph indicating that the two graphs are the same.

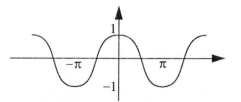

The Graph of an Even Function, Cosine

If f is an odd function, such as $f(x) = x$, the graph is not symmetric about the y-axis.

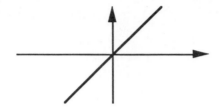

The Graph of an Odd Function, $f(x) = x$

However, it is symmetric with respect to the origin $(0,0)$. That is, if a point (x, y) lies on the graph, so will the point $(-x, -y)$.

The graph of a function that is neither even nor odd is not symmetric about either the y-axis or the origin.

Exercises

1. Give an example of a function, not a polynomial, that is neither even nor odd.

2. Examine functions of the form $f(x) = x^n$ for several values of n. Make a conjecture which of these are even, which are odd. A CAS may make it easier to graph these examples.

3. Extend your exploration of Exercise 2 to other polynomials, e.g., $p(x) = x^3 + 4x^2 - 2x + 1$. What is your conjecture as to which of these are even and which are odd?

4. Extend your exploration of Exercise 2 to trigonometric functions, e.g., $t(x) = \sin^n(x)$. What is your conjecture as to which values of n is t an even function and for which values is t an odd function.

5. Consider $f(x) = \sin(x) - x$. (Note that f is the difference of 2 odd functions.) Is f an odd function? Why? Is the absolute value of f an even function? Are all absolute value functions even functions? Explain.

6. How do even and odd functions combine? For example, is the sum (or difference or product or quotient) of two even functions even? Is the sum (or difference or product or quotient) of an even and an odd function odd? etc.

7. State a conclusion from Exercise 4 as a theorem and prove it.

2.6   Recursive Functions

Prerequisites

- The Getting Started section on the **for** loop

Discussion

Recursive functions are defined in terms of themselves. In order for this to make sense, and avoid infinite regression (where a function infinitely refers to previous undefined values), there must be a *base case* which gets the definition started and an *inductive step* which tells how to define the next value in terms of the current value.

Example 1. The Factorial Function

An example of a recursive definition is the *factorial function*. We define $FACT$ by

$$FACT(0) \ = \ 1 \ \text{(the base case) and}$$

$$FACT(n) \ = \ nFACT(n-1) \ \text{for } n > 0 \ \text{(the inductive step).}$$

Thus

$$FACT(3) = 3 \cdot FACT(2) = 3 \cdot 2 \cdot FACT(1) = 3 \cdot 2 \cdot 1 \cdot FACT(0) = 3 \cdot 2 \cdot 1 \cdot 1 = 6.$$

$FACT(n)$ is often written $n!$. For $n \geq 1$, $n!$ is the product of the first n natural numbers $\triangle$

Example 2. Savings Account

We will determine a recurrence relation giving the amount of money in a savings account after n years when the account pays 6 percent interest and $50 is deposited at the beginning of each year. Let x_k denote the amount at the end of the kth year.

At the end of the nth year there will be the amount at the end of the n-1st year plus $50 plus the interest gained during the nth year. Thus

$$x_0 = 0 \quad \text{(the base case) and}$$

$$x_n \ = \ x_{n-1} + 50 + .06(x_{n-1} + 50) \ = \ 1.06(x_{n-1} + 50) \ = \ 1.06x_{n-1} + 53$$

Hence $x_1 = 53$, $x_2 = 1.06 \cdot 53 + 53 = 109.18$, and $x_3 = 168.73$. $\triangle$

Example 3. Fibonacci's Sequence

We model a rabbit population in which rabbits one year or older are classed as "mature" and those less than one year old are classed as "young." Assume that all pairs of rabbits consist of one male and one female. Suppose every pair of mature rabbits have one pair of offspring every year and that young rabbits have no offspring. The offspring of one year are not counted until the next year, and they become mature (and have their own offspring) the year following that. We will assume that rabbits live indefinitely. Let y_n denote the number of *pairs of young rabbits* in year n, let m_n denote the *number of pairs of mature rabbits* in year n, and let r_n denote the *total number of pairs of rabbits* in year n. We want to determine the sequence $r_1, r_2, r_3, \ldots$.

We first establish some relationships between r_n, m_n, and y_n. Our first relation is "the whole is equal to the sum of the parts":

$$r_n = m_n + y_n \tag{1}$$

Now, any rabbit, young or mature, alive (and counted) in year $n - 1$ will be mature in year n. This gives:

$$m_n = r_{n-1} \tag{2}$$

Also, each pair of young rabbits in year n must be the offspring of a pair of rabbits that were mature in year $n - 1$. Thus:

$$y_n = m_{n-1} = r_{n-2} \tag{3}$$

Substituting for m_n and y_n in the first equation gives:

$$r_n = r_{n-1} + r_{n-2} \tag{4}$$

This relation defines a recursive function, $r(n) = r_n$. In a recursive function, a value is defined by an expression containing earlier values. To prevent infinite regression, some initial values must be given.

Suppose in this example that there is one young pair of rabbits in year 1, so that

$$y_1 = 1, \ m_1 = 0, \ r_1 = 1$$

and

$$y_2 = 0, \ m_2 = 1, \ r_2 = 1$$

Using the recurrence relation (4)

$$r_n = r_{n-1} + r_{n-2}, \ \text{with } r_1 = 1 \text{ and } r_2 = 1$$

we compute

$$r_3 = r_2 + r_1 = 1 + 1 = 2, \quad r_4 = 2 + 1 = 3, \quad r_5 = 5$$

and so on.

The recurrence relation

$$r_n = r_{n-1} + r_{n-2}$$

is called the *Fibonacci Relation* and the resulting sequence r_n (which depends on the initial base conditions, usually taken to be $r_1 = 1$ and $r_2 = 1$) the *Fibonacci Sequence* (named after the Italian mathematician Leonardo of Pisa, called "Fibonacci," 1170-1250). The Fibonacci sequence occurs in many settings in nature and mathematics. An example is the spiral pattern of leaves around a blossom. On an oak or apple stem, there are 5 (r_5) leaves for every two (r_3) spiral turns; on pear stems, 8 (r_6) leaves for every three (r_4) turns; and on willow stems, 13 (r_7) leaves for every five (r_5) spiral turns. There is a mathematics journal, "The Fibonacci Quarterly", devoted to Fibonacci and related recurrence relations. △

We can use the CAS to compute terms of a recursively defined function. To find the first 5 terms of the Fibonacci sequence we would define the first 2 terms and then use a **for** statement to compute the rest:

```
> first:= 1;
> second:= 1;
> for n from 3 to 5 do
>>        next:= first + second;
>>        print(next);
>>        first:= second;
>>        second:= next;
>> od:
```

Note that putting a ":" after **od** will prevent intermediate steps from being printed.

Some computer algebra systems have the capability to find an expression for functions defined recursively. For example, Maple has the command **rsolve** for solving some recursive relations.

Example 4

We will use the **rsolve** command to find an expression for $r(n)$ in the Fibonacci relation

$$r(n) = r(n - 1) + r(n - 2), \qquad r(1) = 1, r(2) = 1$$

we enter (saving the result as "ans")

```
> ans:= rsolve({r(n)=r(n-1)+r(n-2),r(1)=1,r(2)=1}, r(n));
```

and we obtain as output

$$\frac{(1/2 + 1/2\sqrt{5})^n}{\sqrt{5}} - \frac{(1/2 - 1/2\sqrt{5})^n}{\sqrt{5}}$$

The existence of radicals in an expression that must be an integer (rabbits come in whole numbers!) may look unlikely, but we can do some experimentation with it. First we will define a function with this value. Since n is used in the expression, we will use k in the function definition to avoid confusion:

```
> f:= proc(k) subs(n=k, ans) end;
```

Next we test it numerically by printing out the first few values. They probably will require simplification, so we enter

```
> for n from 1 to 5 do simplify(f(n)) od;
```

and we obtain the desired $1, 1, 2, 3, 5$. The base values and first few terms are correct, so we are reasonably confident that this is the correct solution. To make sure, we need to see that f satisfies the recurrence relation $f(n) = f(n-1) + f(n-2)$. We can do the algebra ourselves, or apply the CAS:

```
> simplify(f(k)-f(k-1)-f(k));
```

This may not give the desired result of zero. If it doesn't try some manipulations. In Maple 4.2, the above failed, but simplifying after expansion succeeded:

```
> simplify(expand((f(k)-f(k-1)-f(k))));
```

returned zero.   △

Exercises

1. Compute the first 3 terms of the recursively defined sequence

   $$s(n) = 1 \text{ if } n = 1 \text{ and } s(n) = 2 + s(n-1) \text{ if } n > 1$$

2. Find the 10th term in the Fibonacci sequence, r_{10}.

3. Find the 20th term in the Fibonacci sequence, r_{20}.

4. Consider the recursively defined function given by

   $$s(1) = 2, \ s(2) = -1, \text{ and } s(n) = s(n-1) + s(n-2) \text{ for } n > 2$$

   Find the first 10 terms of s.

5. Does your CAS have a built-in command to solve recursive relations? If it does, try solving the Fibonacci relation, the factorial relation, and the relations of Exercises 1 and 4.

6. Find a recurrence relation for the amount of money in your savings account after n years if the interest rate is 7 percent and you

   (a) deposit $75 at the beginning of each year.

   (b) deposit $75 at the end of each year.

   (c) Evaluate the recurrence relations in parts (b) and (c) when $n = 10$.

7. Suppose you need to carry n chairs from one room into the adjoining room. On each trip you can carry either one or two chairs. Form a recurrence relation for the number of different ways you can carry the n chairs. Evaluate your recurrence relation when $n = 4, 8, 12$.

8. Work Exercise 7 under the assumption that you can carry one, two, or three chairs at a time.

9. Find a recurrence relation for the number of comparisons needed to determine the largest number in a set of n distinct integers. How many comparisons are needed when $n = 5, 8$?

10. Find a recurrence relation for the number of regions determined when n straight lines are drawn on a piece of paper so that every pair of lines intersect, but no three lines intersect in a common point. How many regions are there when $n = 5$?

2.7   Inverse Functions

Prerequisites

- Composition of functions

- The section on Bisection

- The section on Elementary Graphing

Discussion: Existence of Inverse Functions

Let f be a function defined on an interval $[a, b]$. We can consider f to be an "input–output machine" transforming its input, x, to its output $f(x)$. An *inverse function* for f is a function, written f^{-1}, that undoes what f did, transforming the output $f(x)$ back into x. In terms of composition of functions, f^{-1} is the inverse to f on $[a, b]$ if

$$(f^{-1} \circ f)(x) = f^{-1}(f(x)) = x$$

for all x in $[a, b]$. In this section we investigate when a function has an inverse, and how to find the inverse when it exists.

If f has an inverse function, f^{-1}, and $y = f(x)$ then $f^{-1}(y) = x$. Thus, given the equation $y = f(x)$ with x in $[a, b]$, there must be a unique solution for x in terms of y. That is, since a function (f^{-1}) cannot assign two different values to y, there must be only one solution of the equation $y = f(x)$ for x in terms of y. Intuitively, if there are two solutions x and x' to the equation $y = f(x)$, then any inverse function would not know whether to assign to y the value x or the value x'.

Given a function that does not have an inverse we can sometimes restrict the domain of the function and obtain an inverse over the restricted domain.

Example 1

Consider the function $f(x) = (x-1)^2$ on $[0, 2]$. The graph of $y = f(x) = (x-1)^2$ is the parabola:

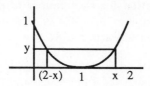

Notice that for any x in $[1,2]$ we have $f(x) = f(2-x) = (x-1)^2$. Thus for any y in $[0,1]$ the equation $y = f(x) = (x-1)^2$ has two corresponding values for x in $[0,2]$, $2-x$ and x. Thus f does not have an inverse function on $[0,2]$.

If we restrict f to the interval $[1,2]$, then f does have an inverse, $f^{-1}(x) = 1 + \sqrt{x}$. Notice that for any x in $[1,2]$,

$$f(f^{-1}(x)) = f(1 + \sqrt{x}) = (\sqrt{x})^2 = x$$

However, for any x in $[0,2]$,

$$f^{-1}(f(x)) = f^{-1}((x-1)^2) = 1 + \sqrt{(x-1)^2} = 1 + |x-1|$$

For example, $f^{-1}(f(0)) = 2$.
Warning: many CASs, e.g., Maple and Mathematica, assume that $1 + \sqrt{(x-1)^2} = x$. If we define these functions with one of these CASs

```
> f:= proc(x) (x-1)^2 end;
> finv := proc(x) 1 + sqrt(x-1) end;
```

Then

```
finv(f(0));
```

returns 2, but for an unassigned variable x,

```
finv(f(x));
```

returns x. Not all CASs assume that $\sqrt{(x-1)^2} = x - 1$: MACSYMA returns $\text{abs}(x-1)$.   △

From this example we see that if the graph of a continuous function changes direction in an interval, then there can be no inverse function in that interval. However if the graph of a continuous function does not change direction over an interval, then there is an inverse function.

Theorem 2.7.1 *If f is strictly increasing on an interval $[a,b]$, then f has an inverse function on that interval.*

Proof: We need to show that there cannot be two solutions to the equation $y = f(x)$. Suppose $y = f(x_1)$ and $y = f(x_2)$. One of $x_1 = x_2$, $x_1 < x_2$, or $x_1 > x_2$ must hold. If $x_1 < x_2$, then since f is strictly increasing, $y = f(x_1) < f(x_2) = y$. This is impossible, as is the case $x_1 > x_2$. Thus $x_1 = x_2$ and there cannot be more than one solution to the equation.   □

Note from this proof that if f is strictly increasing, then so is f^{-1}. A similar result holds for functions that are strictly decreasing on an interval.

Example 2

Suppose that a function f has the following graph on the interval $[a, e]$.

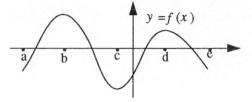

We see that f does not have an inverse on $[a, e]$ since, for example, the equation $0 = f(x)$ has 4 solutions for x in $[a, e]$. However, on each of the subintervals $[a, b]$, $[b, c]$, $[c, d]$, and $[d, e]$, the function f is either strictly increasing (on $[a, b]$ and $[c, d]$) or strictly decreasing (on $[b, c]$ and $[d, e]$). Thus on each of these subintervals f has an inverse function. △

Example 3

Let $f(x) = \sin(x)$, which is defined for all real numbers.

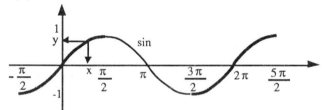

On each subinterval $[-\frac{\pi}{2}, \frac{\pi}{2}]$, $[\frac{\pi}{2}, \frac{3\pi}{2}]$, ..., f has an inverse function. Texts usually choose the interval $[-\frac{\pi}{2}, \frac{\pi}{2}]$ and define $\arcsin(x) = \sin^{-1}(x)$ on this interval.

In terms of this arcsin, what is the inverse of sin restricted to $[\frac{3\pi}{2}, \frac{5\pi}{2}]$? For any y value in $[-1, 1]$, the x value of $\arcsin(y)$ must be shifted horizontally by 2π. Thus if f is sin restricted to $[\frac{3\pi}{2}, \frac{5\pi}{2}]$, $x = f^{-1}(y) = \arcsin(y) + 2\pi$.

In terms of this standard arcsin, what is the inverse of $f = \sin$ restricted to $[\frac{\pi}{2}, \frac{3\pi}{2}]$? This is more difficult, since when $y = 1$, $f^{-1}(y) = \arcsin(y)$ and when $y = -1$, $f^{-1}(y) = \arcsin(y) + 2\pi$. We want to reflect in the line $x = \frac{\pi}{2}$ so that (x, y) becomes $(\frac{\pi}{2} - x, y)$. Thus $f^{-1}(y) = \frac{\pi}{2} - \arcsin(y)$. △

Exercises

1. Estimate, using the graph, the subintervals of $[-2, 2]$ where $f(x) = 3x^3 - 6x^2 - 3x + 6$ has an inverse.

2. Estimate, using the graph, the subintervals of $[-2, 2]$ where $f(x) = 4x^4 - 4x^3 - 16x^2 + 16x$ has an inverse.

3. Estimate, using the graph, the subintervals of $[-2, 2]$ where $f(x) = 1$ has an inverse.

4. Estimate, using the graph, the subintervals of $[0.1, 1]$ where $f(x) = \sin(\frac{1}{x})$ has an inverse.

5. Let arcsin be the usual inverse of sin restricted to the interval $[-\frac{\pi}{2}, \frac{\pi}{2}]$. In terms of arcsin, what is the inverse of sin restricted to $[-\frac{5\pi}{2}, -\frac{3\pi}{2}]$?

6. Let arcsin be the usual inverse of sin restricted to the interval $[-\frac{\pi}{2}, \frac{\pi}{2}]$. In terms of arcsin, what is the inverse of sin restricted to $[-\frac{3\pi}{2}, -\frac{\pi}{2}]$?

Discussion: Computation of Inverse Functions

Suppose that a function f has an inverse on $[a, b]$. How can f^{-1} be found? If $f(x) = y$, then $f^{-1}(y) = x$. Thus we can find f^{-1} by solving the equation $f(x) = y$ for x in terms of y. Sometimes this is easy to do.

Example 4

Since $f(x) = (x - 1)^4$ is strictly increasing on the interval $[1, 3]$, it has an inverse on this interval. We can solve $f(x) = (x - 1)^4 = y$ for x in terms of y easily enough:

$$
\begin{aligned}
(x - 1)^4 &= y \\
(x - 1) &= \pm y^{1/4} \\
x &= 1 \pm y^{1/4}
\end{aligned}
$$

We have two solutions, since our solution method did not distinguish the interval involved. Since f is increasing on $[1, 3]$, f^{-1} is also increasing, so we have

$$
f^{-1}(y) = 1 + y^{1/4}
$$

We can try the `solve` command of our CAS:

```
> solve((x-1)^4 = y,x);
```

returns

$$
1 + y^{1/4}, \quad 1 - y^{1/4}, \quad 1 + Iy^{1/4}, \quad 1 - Iy^{1/4}
$$

Where I represents the complex number $i = \sqrt{-1}$. Note that we get the complex solutions as well as the real solutions. From the solutions returned by the CAS we must use our knowledge of the behavior of f (and thus f^{-1}) on the interval in question to pick out the correct solution, $f^{-1}(y) = 1 + y^{1/4}$. $\triangle$

Example 5

We will find the inverse function to $f(x) = \sin(x^2 + 1) - 2$ on $[2, 2.25]$. First we define and graph f on this interval:

```
> f:= proc(x) sin(x^2 + 1) - 2 end;
> plot(f(x),x=2..2.25);
```

This gives

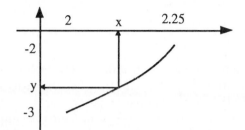

We see that f is strictly increasing on $[2, 2.25]$, so an inverse exists. We can solve by hand, or use the command

```
> solve(f(x)=y,x);
```

which returns

$$-(\arcsin(y + 2) - 1)^{1/2}, \quad (\arcsin(y + 2) - 1)^{1/2}$$

Since x is positive, we want the positive root. However, as in Example 3, we know that arcsin is the standard inverse for sin over the interval $[-\frac{\pi}{2}, \frac{\pi}{2}]$ and may not be the desired inverse. In fact, if $y \approx -2$ then $\arcsin(y + 2) - 1 < 0$ and no real square root exists. Thus we need to replace arcsin by the inverse of sin when $x^2 + 1$ is near $2^2 + 1 = 5$. As in Example 3 the appropriate inverse is $\arcsin + 2\pi$. So our inverse should be $(\arcsin(y + 2) + 2\pi - 1)^{1/2}$. We can check by defining this function

```
> finv := proc(y) sqrt(arcsin(y+2)+2*Pi - 1) end;
```

and evaluating at a few points, $f(finv(-2.8)) = -2.8$ and $finv(f(2)) = 2$. Warning: some CASs, e.g., Maple and MACSYMA, requre

```
> evalf(finv(evalf(f(2))));
```

rather than just `evalf(finv(f(2)))`.   △

Exercises

7. Find the inverse function to $f(x) = 1 + (x + 2)^2$ on $[-4, -3]$, if it exists.

8. Find the inverse function to $f(x) = \sqrt{x^2 + x + 1}$ on $[-5, -2]$, if it exists.

9. Find the inverse function to $f(x) = \cos(2x)$ on $[-\frac{\pi}{4}, 0]$, if it exists.

10. Find the inverse function to $f(x) = 3 + \sin(x^4 + 1)$ on $[1.7, 1.75]$, if it exists.

2.8   Approximation and Error Bounds

Prerequisites

- Absolute values and inequalities

- The section on Roots (Zeros) of Polynomials

- The section on Data Fitting

Discussion

The process of approximation is a central theme in calculus. It gives a solution to the problem of computing difficult quantities: find an easily computed quantity which is sufficiently close to the desired quantity.

In the development of concepts as well as in numerous applications, a crucial step often involves approximating a given expression to within a stated accuracy. As you know, a function is a rule that assigns a definite value $f(x)$ to each value x in the domain of f. To find the value of $f(x)$ exactly, we must know x exactly. This point seems trivial until we realize that in many situations we have only approximations for x available! This reality is often the result of imperfections in measuring devices and other data-gathering mechanisms. More importantly, the necessity of approximation is an artifact of the number system and cannot be avoided. For example, the diagonal of the unit square has length $\sqrt{2}$. Although there exists an algorithm for computing the decimal expansion of the square root of two, it requires an infinite number of operations! Thus numerical expressions for $\sqrt{2}$ are, by necessity, approximations.

Every real number has an infinite decimal representation. Rational numbers are characterized as real numbers whose decimal expansions are eventually repeating. Irrational numbers are real numbers whose decimal expansions are non-repeating (such as $\sqrt{2}$). Decimal expressions for all irrational numbers and for most rational numbers are approximations. For example, 1.414 is an approximation to $\sqrt{2}$. Even when not working with irrational numbers, many of the numerical printouts of your calculator are approximations, since the calculator only works with a limited number of digits of accuracy (usually about 8).

Approximations are also used in working with symbolic expressions. For example, for "small" values of x the expression $y = \sin(x)$ is often approximated by $y = x$ when x is near zero. This means that when the value of x is near zero, the value of $\sin(x)$ is near the value of x. Another example is approximating a portion of a rational function by an asymptote. For example, for "large" values of x the expression $y = \frac{x^2+1}{x}$ is approximated by $y = x$.

A basic question associated with any approximation is: How good is the approximation? That is, what is the error? Of course, no exact numerical description of the error can be given (otherwise there would be no need to use an approximation). Thus we introduce the term "error bound," an upper bound on the size of the error. It is important to realize that although the absolute value of the error may be considerably smaller than the error bound, it can never be larger. In general, the smaller the error bound the better the approximation. Accuracy, abbreviated ACC (or by the Greek letter ϵ), is often used as a synonym for error bound.

Example 1. Approximation of $\frac{45}{53}$

We will approximate $\frac{45}{53}$ with $ACC = 10^{-3}$.

Using a computer, calculator, or long division, we find the decimal expansion for $\frac{45}{53}$.

$$\frac{45}{53} = 0.84905660377...$$

The desired approximation is 0.849. Note that the error is less than 10^{-3}. That is,

$$error = |\frac{45}{53} - 0.849| = 0.00005660377... < 10^{-3}$$

Actually the error in the above approximation is less than 10^{-4}. $\triangle$

Example 2. Approximation of $x\sin(x)$ by x^2

We will determine an interval over which x^2 approximates $x\sin(x)$ with an accuracy of 0.1 or less. That is, determine an interval over which $|x\sin(x) - x^2| < 0.1$. We first transform this problem into one of finding the zeros of a function and then use a graphical approach to approximate the zeros.

Since $|x\sin(x) - x^2| < 0.1$ is equivalent to $|x\sin(x) - x^2| - 0.1 < 0$, we define a function f by $f(x) = |x\sin(x) - x^2| - 0.1$ and determine the interval over which $f(x)$ is negative. This is done by plotting f and determining (approximating) the zeros.

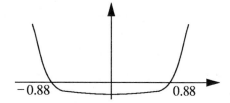

-0.88 0.88

Using the methods illustrated in the section on Roots(Zeros) of Polynomials, we approximate the zeros to be -0.88 and 0.88. Thus x^2 approximates $x\sin(x)$ over the interval $[-0.88, 0.88]$ with an accuracy of 0.1. $\triangle$

Example 3. Approximation by a Polynomial

In this example we will find a quadratic polynomial that approximates the cosine function over the interval $[-\frac{\pi}{2}, \frac{\pi}{2}]$, and will estimate the error in the approximation. We define a general quadratic polynomial function:

```
> p:= proc(x) a*x^2 + b*x + c end;
```

We need three equations to solve for a, b, and c. To obtain them we will set p equal to cos at three points, $x = -\frac{\pi}{2}$, $x = 0$, and $x = \frac{\pi}{2}$ and then apply the solve command to solve this *system* of equations for a, b, and c:

```
> solve({p(-Pi/2)=cos(-Pi/2),p(0) = cos(0),
≫              p(Pi/2) = cos(Pi/2)},{a,b,c});
```

$$\{a = -\frac{4}{\pi^2}, \ b = 0, \ c = 1\}$$

(As with other commands, solve will work on a set of expressions, returning a set as an answer. Notice that in Maple's notation, if the order of items does not matter, then set notation is used: {a,b c}. If order *does* matter, then list notation is used: [a,b,c].) Now we substitute these values into p giving the desired polynomial which we call q:

```
> q:= subs(",p(x));
```

$$q := -\frac{4x^2}{\pi^2} + 1$$

(Notice that p is a function; $p(x)$ is p evaluated at x, an expression. The result of substituting values for a, b, and c in the expression $p(x)$ is the expression q.) We can estimate the error by graphing the difference between cos and q over the interval $[-\frac{\pi}{2}, \frac{\pi}{2}]$:

```
> plot(cos(x)-q,x=-Pi/2..Pi/2);
```

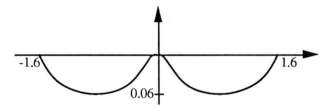

We can see that the error is at most 0.06.   △

Exercises

1. Prove or disprove that 10^{-4} is an error bound when $\frac{33}{63}$ is used to approximate 0.660714282571 ....

2. Find a decimal approximation to $\frac{21}{63}$ with an accuracy of 10^{-6}.

3. Sketch the graphs of $y = \sin(x)$ and $y = x$ on the same set of axes. Determine an interval centered at $x = 0$ over which $y = x$ approximates $y = \sin(x)$ with accuracy 0.1.

4. Sketch the graphs of $y = \dfrac{x^2 + 1}{x}$ and $y = x$ on the same set of axes. How large must x be in order that $y = x$ approximates $y = \dfrac{x^2 + 1}{x}$ with accuracy 1?

5. Determine the best error bound that you can when $y = \sin(x)$ is approximated by $y = x$ over $[-\frac{\pi}{2}, \frac{\pi}{2}]$. Hint: Use a calculator, computer, or set of tables to compute $|x - \sin(x)|$ for various values of x.

6. Estimate the error when the sine function is approximated over the interval $[-\frac{\pi}{2}, \frac{\pi}{2}]$ by

   (a) a quadratic polynomial

   (b) a cubic polynomial

7. Estimate the error when the cosine function is approximated over the interval $[-\frac{\pi}{2}, \frac{\pi}{2}]$ by

   (a) a quadratic polynomial

   (b) a cubic polynomial

8. Estimate the error when the tangent function is approximated over the interval $[-\frac{\pi}{3}, \frac{\pi}{3}]$ by

   (a) a quadratic polynomial

   (b) a cubic polynomial

9. Determine whether $y = x^2$ or $y = x^3$ is a better (i.e., has a smaller error bound) approximation to $y = \tan(x)$ over the interval $[0, 1]$.

10. Consider the true statement: If a and b have the same integer part and agree up to the mth decimal place, then they differ by less than 10^{-m}.

    (a) Produce a counterexample to show that the converse of the above statement is false. (That is, show that: If a and b differ by less than 10^{-m}, then they have the same integer part and agree up to the mth decimal place.)

    (b) Let p be an integer less than 10 and let a and b be any numbers on opposite sides of $p10^{-m}$. Show that the mth digits of a and b are different (no matter how close a and b are together).

    (c) Develop a theorem (and proof) that tells how close two numbers are by looking at the digits in their decimal expansion. Hint: Consider the "rounded up" and "rounded down" expression after m digits for each of the two numbers.

2.9   Convergence of Sequences

Prerequisites

- An elementary knowledge of sequences

- The section on Approximation and Error Bounds

Discussion

Convergence is the primary concept that distinguishes calculus from algebra. The derivative and integral, which generate the major branches of calculus, are both defined in terms of convergence.

The question of convergence is the question of obtaining an approximation to any specified accuracy (i.e., error bound). We illustrate the idea of convergence by considering the following sequence of approximations to $\frac{1}{3}$. Let

$$s_1 = 0.3$$

$$s_2 = 0.33$$

$$s_3 = 0.333$$

$$\vdots$$

$$s_n = 0.333 \cdots 3 \quad (n \text{ threes})$$

The error associated with s_1 is $|s_1 - \frac{1}{3}| = |0.3 - \frac{1}{3}| = \frac{1}{3 \cdot 10}$.

The error associated with s_2 is $|s_2 - \frac{1}{3}| = |0.33 - \frac{1}{3}| = \frac{1}{3 \cdot 10^2}$.

$$\vdots$$

The error associated with s_n is $|s_n - \frac{1}{3}| = \frac{1}{3 \cdot 10^n}$.

Note that the error can be made as small as desired by choosing a sufficiently large value for n. Thus given any accuracy, a real number $ACC > 0$, there exists an n such that $|s_n - \frac{1}{3}| < ACC$. For example, if $ACC = \frac{7}{1000}$, solving $|s_n - \frac{1}{3}| < ACC$ for n yields

$$\frac{1}{3 \cdot 10^n} < \frac{7}{1000} \quad \Longleftrightarrow$$

$$\frac{1}{3 \cdot 10^{n-3}} < 7 \quad \Longleftrightarrow$$

$$\frac{1}{21} < 10^{n-3} \quad \Rightarrow \quad n \geq 2$$

Can your CAS solve inequalities? The above is a good test. In Maple, we enter

```
> evalf(solve(1/(3*10^n)<7/1000,n));
```

$$1.677... < n$$

Since n must be an integer, we have $n \geq 2$. (Notice that we used **evalf** to obtain a decimal answer, rather than an expression involving logarithms.)

Since for any accuracy, ACC, there exists an N such that for any $n \geq N$, $|s_n - \frac{1}{3}| < ACC$, we say the sequence of approximations $\{s_n\}$ converges to $\frac{1}{3}$.

Definition 2.9.1 *A sequence $\{s_n\}$ converges to L if, for any accuracy, $ACC > 0$, there exists a number N such that $|s_n - L| < ACC$ for all $n > N$. N may depend on the value of ACC.*

L is called the limit of the sequence. If a sequence does not converge, it is said to *diverge*. The following three expressions are all equivalent.

(1) $\{s_n\}$ converges to the number L

(2) $\lim\limits_{n \to \infty} s_n = L$

(3) $\{s_n\}$ approximates the point L for all ACC.

Since a sequence is a function defined only on the positive integers, its graph is a set of points rather than a continuous curve. Thus in plotting a sequence we shall often use a *point* plot (in contrast to a line or a curve plot). We illustrate with a point plot of $\{\frac{1}{n}\}$, one of the simplest but most important sequences in calculus.

Example 1

We form a point plot of $\{\frac{1}{n}\}$ over $[1, 50]$. The Maple command is

```
> plot(1/n, 1..50, style=POINT);
```

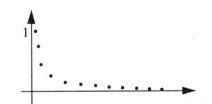

This point plot *strongly suggests* that $\{\frac{1}{n}\}$ is a bounded, monotonically decreasing sequence converging to zero, although it does not prove this result.   △

The next example illustrates how a point plot can be used to help confirm a conjectured value for the limit of a sequence.

Example 2

Consider the sequence $\left\{ \dfrac{n^3 - 12n}{2n^3 + 10} \right\}$. Since both the numerator and denominator will behave like their highest powered terms for large values of n, we expect that this sequence will behave like $\left\{ \dfrac{n^3}{2n^3} \right\}$ or $1/2$ for large values of n. Thus we conjecture that the limit is $1/2$. To "test" our conjecture, we draw a multipoint plot of the sequence and the conjectured limit.

```
> plot({(n∧3-12*n)/(2*n∧3+10),1/2},1..25, style=POINT);
```

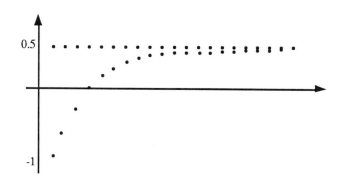

Although not a proof our plot strongly suggests that the limit is $1/2$.   △

Sometimes it is possible to determine that a sequence converges without knowing the limit.

A sequence is *monotonic* if it is either always increasing or always decreasing. A monotonic sequence can stay at the same value, but it cannot reverse. Algebraically, $\{s_n\}$ is monotonically increasing if and only if $s_m \leq s_n$ whenever $m \leq n$. For example, $\{\frac{1}{n}\}$ steadily decreases, so it is monotonically decreasing. The corresponding negative sequence, $\{-\frac{1}{n}\}$, is monotonically increasing.

A sequence $\{s_n\}$ is bounded if there is a positive number B (a bound on $\{s_n\}$) such that $|s_n| \leq B$ for all n. For example, the sequence $\{\frac{1}{n}\}$ is bounded (by $B = 1$) since $|s_n| = \frac{1}{n} \leq 1$ for all $n \geq 1$.

Theorem 2.9.2 *If a sequence is bounded and monotonic, then it converges.*

Since the sequence $\{\frac{1}{n}\}$ is both bounded and monotonic, we know that it converges. In this case, it is easy to see that it converges to zero. (We will go through the details, using the definition of convergence, in the first example below.)

Example 3

We will show that the sequence $\{\frac{1}{n}\}$ converges to 0 by showing that the conditions of the definition are satisfied. We know from the above theorem that it converges, but the theorem does not tell us the limit. Let ACC be any positive number. We need to show that there is a number N such that $|\frac{1}{n} - 0| < ACC$ for all $n > N$. Since

$$|\frac{1}{n} - 0| < ACC \Longleftrightarrow \frac{1}{n} < ACC \Longleftrightarrow \frac{1}{ACC} < n$$

we can let N be any integer greater than $\frac{1}{ACC}$. Now for any $n > N$

$$|\frac{1}{n} - 0| = \frac{1}{n} \leq \frac{1}{N} < ACC$$

Thus $\{\frac{1}{n}\}$ converges to zero.   △

Example 4

In this example we will not *prove* that a sequence converges to a particular value but will use our CAS to *determine* the limit, $\lim\limits_{n \to \infty} \dfrac{n^3 - 4n^2 + 2}{2n^3 - 7}$. We first define the expression on our CAS

```
> s:=(n∧3 - 4*n∧2 +2)/(2*n∧3 - 7);
```

To evaluate the limit of s as n goes to infinity, we write

```
> limit(s, n = infinity);
```

The CAS responds with: $\frac{1}{2}$.   △

Example 5

We will evaluate $\lim\limits_{n \to \infty} \dfrac{1 + 2 + 3 + \cdots + n}{n^3 + 17}$.

This is a two step problem. The first step is to sum the numerator and the second step is to evaluate the limit of the resulting rational expression (i.e., the problem of Example 2). We sum the numerator and assign it the name s.

```
> s:=sum(i,i=1..n);
```

The response is:

$$\frac{1}{2}(n + 1)^2 - \frac{1}{2}n - \frac{1}{2}$$

(Note that this is the correct answer, but it is usually written as $\frac{n(n+1)}{2}$. A CAS will sometimes return the correct answer in an unexpected form.) We could carry out step 2 by first defining a function and then taking the limit of that function as was done in Example 2. Another way is to evaluate the limit of the rational expression directly, using a CAS limit command.

```
> limit(s/(n∧3+17),n=infinity);
```

The result is 0.   △

Example 6

We will evaluate $\lim\limits_{n\to\infty}\dfrac{n^4+3n^2+5}{6n^3+2n+1}$. Using our CAS, we have:

```
> limit((n∧4 + 3*n∧2 + 5)/(6*n∧3 + 2*n + 1),n = infinity);
```

$$\text{infinity}$$

In this case the limit does not exist (i.e., the sequence diverges). Why? When n is large, the numerator is dominated by the first term, n^4, and the denominator is dominated by the term $6n^3$. The ratio of these is $n^4/6n^3 = n/6$ which goes to infinity as n goes to infinity. $\triangle$

The following example illustrates using the definition of convergence to solve a typical approximation problem when the limit of a function is known.

Example 7

Given that the sequence $s_n = \dfrac{2n^2+17}{n^2+n}$ converges to 2, we will determine a number N such that s_n approximates 2 with accuracy 0.01 for all $n > N$.

The problem is to determine an N such that for $n > N$,

$$|s_n - 2| = \left|\frac{2n^2+17}{n^2+n} - 2\right| < 0.01$$

We will solve this problem in three ways. The first two will use the graphic capabilities of a CAS and the third will use the numerical and symbolic capabilities of a CAS. First we define our function s:

```
> s:= proc(n) (2*n∧2 + 17)/(n∧2 + n) end;
```

Graphical Solution, Method A. We begin by expressing the absolute value, inequality expression $|s_n - 2| < 0.01$ as the double inequality $2 - 0.01 < s_n < 2 + 0.01$. The problem is to determine a value of N such that this double inequality is satisfied for all $n > N$. Geometrically this means to determine an N such that the graph of s_n lies between the lines $y = 1.99$ and $y = 2.01$ for all $n > N$. Thus we draw a multiplot of these three curves. We start with a large interval and then zoom in on the point where s_n crosses the line $y = 2-0.01$. This may require several applications of zooming. For example, consider plotting over the interval $1 \le n \le 1000$, then over $50 \le n \le 200$, then over $100 \le n \le 200$, then over $150 \le n \le 200$, etc.

```
> plot({sₙ, 2 − 0.01, 2 + 0.01}, n = 150..200);
```

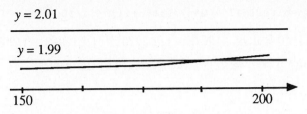

Continuing to plot over smaller intervals, we approximate a solution of $N = 192$.

Graphical Solution, Method B. We transpose the absolute value, inequality problem $|s_n - 2| < 0.01$ into a problem of finding the zeros of a function. Define a function g by $g(n) = |s_n - 2| - 0.01$. Our problem is now to determine an N such that $g(n) < 0$ for all $n > N$. We will do this by plotting $g(n)$ and letting N be the smallest integer for which the graph of g lies under the horizontal axis for all $n > N$. It appears that $g(n) < 0$ for all $n \geq 191$, but how can we be sure? Although we can continue to plot over larger and larger intervals, we will eventually have to resort to an analytical argument to show that $N = 191$ is correct. (Note that $g(n) < 0$ for n close to 8.)

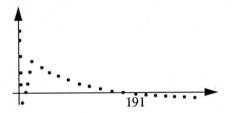

As with Method A, the intervals over which g is graphed may have to be adjusted several times before a plot can be obtained that enables a sufficiently accurate approximation for the number N.

Numerical Solution. Since solving an absolute value, inequality expression is too difficult for many CASs, we apply the definition of absolute value to $|s_n - 2|$ in an attempt to simplify the problem to solving just an inequality expression.

```
> simplify(s(n) - 2);
```

$$-\frac{-17 + 2n}{n(n + 1)}$$

Since this expression is negative for $n > 9$, we have

$$|s_n - 2| = -(s_n - 2) = \frac{-17 + 2n}{n(n + 1)} \quad \text{for} \quad n > 9$$

We now restate our problem as: Determine an $N > 9$ such that $\dfrac{-17 + 2n}{n(n + 1)} < \dfrac{1}{100}$ for all $n > N$. We use a CAS to solve this inequality, assigning the name r to the solution set.

```
> r:= solve((-17 + 2*n)/(n*(n+1)) < 1/100,n);
```

$$r := \{n < -1\}, \{0 < n, n < 199/2 - 1/2\ 32801^{1/2}\}, \{199/2 + 1/2\ 32801^{1/2} < n\}$$

Since we require $n > 9$, we select the third set in the list r.

```
> evalf(r[3]);
```

$$\{190.055231 < n\}$$

Thus our solution is $N = 191$.

$\triangle$

Exercises

Determine if the sequences in Exercises 1 through 3 converge or diverge using first a graphical approach (i.e., point plot) and then an analytical approach.. If a sequence converges, evaluate the limit. If a sequence diverges, explain why it diverges.

1. $x_n = \dfrac{4n^2 + 7n - 1}{n^2 + 4}$

2. $x_n = \dfrac{3n^5 + 13n^2 - 8n}{14n^9 - 7n^6 + 13n^5 - 17}$

3. $x_n = \dfrac{2n^4 - 16n}{n^2 + 2}$

4. Based on the results of Exercises 1, 2, and 3, make a conjecture for

$$\lim_{n \to \infty} \frac{an^t + (additional\ terms)}{bn^s + (additional\ terms)}$$

where t is the largest exponent in the numerator and s is the largest exponent in the denominator. Hint: If a pattern of behavior is not clear from Exercises 1, 2, and 3, make up and work additional problems similar to Exercises 1 through 3.

5. Evaluate $\displaystyle\lim_{n \to \infty} \frac{1 + 4 + 9 + \cdots + n^2}{n^3 + n}$.

6. Evaluate $\displaystyle\lim_{n \to \infty} \frac{(1 + 4 + \cdots + n^2) - (1 + 2 + \cdots + n)}{n(n - 5)}$.

7. Evaluate $\displaystyle\lim_{n \to \infty} \frac{1 + 16 + 81 + \cdots + n^4}{(1 + 2 + 3 + \cdots + n)(1 + 4 + 9 + \cdots + n^2)}$.

8. Determine a number N such that for $n > N$, $s_n = \dfrac{3n^3 + 13}{n^3 - 12}$ approximates 3 with accuracy 0.01.

9. Determine a number N such that for $n > N$, $s_n = \dfrac{2n^3 - 13}{n^2(n+1)}$ approximates 2 with accuracy 0.01.

10. Consider the sequence $A_n = \dfrac{(2)(4)(6)\cdots(2n)}{(1)(3)(5)\cdots(2n-1)}$.

   (a) Show algebraically that $A_n = \dfrac{(2^n n!)^2}{(2n)!}$.

   (b) Observe several values of $\sqrt{A_n}/n$ for n between 1 and 1,000 (say, $n = 9,16,25,75,100,150,196$, etc.).

   (c) Using the results of part (b) as a guide, apply algebra to find upper and lower bounds on $\sqrt{A_n}/n$, for all $n \geq 1$.

   (d) Show by algebra that $\sqrt{A_n}/n$ is a monotonic sequence and thus, since it is bounded, is convergent.

   (e) Conjecture what the limit of $\sqrt{A_n}/n$ is by observing several values of $\sqrt{A_n}/n$ for n between 500 and 1,000.

2.10   Limits of Functions

Prerequisites

- Basic function concepts

- The Section on Approximation and Error Bounds

Discussion: Part 1

A function may be considered as an input – output relation, where x denotes the input and $f(x)$ the output. The basic idea of limit is that as x gets closer and closer to a number c, $f(x)$ gets closer and closer to some number, say L. This description, while possessing intuitive clarity, lacks the precision necessary for mathematical use (e.g., what does "closer and closer" mean?). We can, however, give this statement a precise and formal meaning by describing how the distances $|x - c|$ and $|f(x) - L|$ are related through the functional relation. Before stating a precise description, we illustrate the ideas involved by an example.

Example 1

When x gets close to 3, $f(x) = x^2 - 3x + 2$ gets close to 2. How close must x be to 3 in order to guarantee that $f(x)$ will be within 0.001 of 2? That is, how small must $|x - 3|$ be in order to have $|f(x) - 2| < 0.001$? (Note that $|f(x) - 2| < 0.001$ means that $1.999 < f(x) < 2.001$.)

We will illustrate three solution methods, the first two will be graphical and the third one analytical.

Method 1

Since we want the values of x near 3 for which $1.999 < f(x) < 2.001$, we draw a multiplot of f over a small interval about 3, say [2.5,3.5], and the two lines $g(x) = 2.001$ and $h(x) = 1.999$.

```
> plot({f(x),g(x),h(x)}, x=2.5..3.5);
```

By graphing over increasingly smaller intervals, we obtain a graph similar to the following:

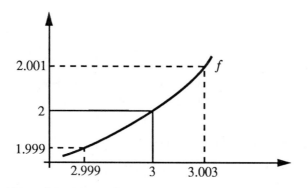

The values 2.999 and 3.003 are the x coordinates of the points of intersection of the lines $g(x) = 2.001$ and $h(x) = 1.999$ with the graph of f. These values can be obtained algebraically by solving the equations $f(x) = 1.999$ and $f(x) = 2.001$ or by digitizing the intersection points on the graph.

Although $f(x)$ is within 0.001 of 2 for any x in the interval (2.999, 3.003), this does not the answer since the question asked *how close* must x be to 2. Thus we compute the $\min(|3 - 2.999|, |3 - 3.003|)$. Since $\min(|3 - 2.999|, |3 - 3.003|)$ = .001, we know that if x is within 0.001 of 3 then $f(x)$ will be within 0.001 of 2. (That is, $|x - 3| < 0.001$.)

Method 2

We will convert the problem into one of finding the zeros of a function. We want $|f(x) - 2| < 0.001$ which is equivalent to $|f(x) - 2| - 0.001 < 0$. Thus we define the function g by $g(x) = |f(x) - 2| - 0.001$ and determine the values of x near 3 for which $g(x) < 0$. We do this by plotting g and approximating the zeros (x intercepts).

```
> plot({f(x),g(x),h(x)},x=2.5..3.5);
```

By graphing over increasingly smaller intervals, we obtain a graph similar to the following:

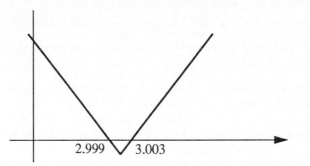

2.999 3.003

This is the same result that we had obtained in the previous method. Thus we want x to be within 0.001 of 3 for $f(x)$ to be within 0.001 of 2. (That is, $|x - 3| < 0.001$.)

Method 3

We will now determine an answer analytically (without referring to a graph). That is, we will determine the values of x near 3 that satisfy the absolute value-inequality expression

$$|f(x) - 2| = |(x^2 - 3x + 2) - 2| < .001$$

Unfortunately most CASs will fail on the problem

```
> solve(abs((x^2 - 3*x + 2)-2) < 0.001,x);
```

Thus, even with a CAS one must sometimes do the algebra (or part of it) oneself. The following sequence of algebraic steps will lead us to an answer.

$$
\begin{aligned}
|f(x) - 2| < 0.001 \quad &\Longleftrightarrow \quad |(x^2 - 3x + 2) - 2| < 0.001 \\
&\Longleftrightarrow \quad |x^2 - 3x| < 0.001 \\
&\Longleftrightarrow \quad |x||x - 3| < 0.001 \\
&\Longleftrightarrow \quad |x - 3| < \frac{0.001}{|x|}
\end{aligned}
$$

In order to determine a numerical bound for the right-hand side we note that since x is approaching 3, we may restrict x to the interval [2,4]. Then for $x \geq 2$

$$\frac{0.001}{|x|} \leq \frac{0.001}{2}$$

and thus

$$|f(x) - 2| < 0.0001 \quad \text{if} \quad |x - 3| < \frac{0.001}{|x|} \leq \frac{0.001}{2}$$

Thus $f(x)$ will be within 0.001 of 2 whenever x is within $\frac{0.001}{2}$ of 3. $\qquad \triangle$

Note that restricting x to a different interval about $x = 2$ could yield a different numerical bound on $\dfrac{0.001}{|x|}$ which in turn could give a different answer. In general, the analytical approach will not give the largest possible answers for x (as does the graphical approach).

It is important to understand how the restriction on how close $f(x)$ must be to 2 determined a corresponding restriction on how close x must be to 3. These restrictions are called *error bounds* for $f(x)$ and x, respectively. It is traditional to use the Greek letters ϵ and δ to denote the error bounds for $f(x)$ and x, respectively. Thus in this example if ϵ had been given as 0.000001 (instead of 0.001), a corresponding value for δ would have been 0.000001/2. Or, if ϵ was given any positive value, a corresponding value for δ would be $\epsilon/2$. This last statement expresses the "heart" of the limit concept. That is, given any (positive) error bound ϵ for $|f(x) - 2|$ there exists a (positive) error bound δ for the $|x - 3|$.

Definition 2.10.1 *The limit of $f(x)$ as x approaches c is a number L, written $\lim\limits_{x \to c} f(x) = L$, provided that given any positive number ϵ there is a corresponding positive number δ such that $|f(x) - L| < \epsilon$ whenever $0 < |x - c| < \delta$.*

If no δ value exists for some given ϵ value, then the limit of the function does not exist.

Graphically the definition says that the oscillation of $f(x)$ about the limit L (i.e., $|f(x) - L| < \epsilon$) can be made arbitrarily small by restricting x to be sufficiently close to c (i.e., $0 < |x - c| < \delta$).

Example 2

We will see that $\lim\limits_{x \to 1} \dfrac{1}{1 - x}$ does not exist, using several different approaches.

First, we can ask our CAS:

```
> limit(1/(1-x), x=1);
```

Returns

$$\text{undefined}$$

The result can also be seen by looking at a graph of $y = f(x) = \frac{1}{1-x}$ (below) since y is unbounded above as x approaches 1 from the left and y is unbounded below as x approaches 1 from the right. These observations can be verified by using a CAS to compute the left- and right-hand limits.

```
> limit(1/(1-x), x=1, left);
```

$$\text{infinity}$$

```
> limit(1/(1-x), x=1, right);
```

−infinity

Thus the oscillation of y is infinite over any open interval about $x = 1$.

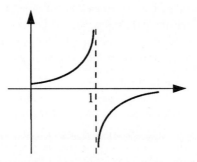

Intuitively it is clear that the limit does not exist, since when x is near 1 the numerator is 1 and the denominator is near 0. Thus the magnitude of the ratio is very large (either positive or negative). Hence the limit does not exist.

To establish the result analytically, where we show that no number L can satisfy the definition, we note that if $\dfrac{1}{2} < x < 1$, then $1 - x < 1 - \dfrac{1}{2}$ and so $\dfrac{1}{1-x} > \dfrac{1}{1-1/2} = 2$. Similarly, if $1 < x < \dfrac{3}{2}$, then $\dfrac{1}{1-x} < \dfrac{1}{1-3/2} = -2$. Thus for $\dfrac{1}{2} < x < 1 < z < \dfrac{3}{2}$, we have $|f(x) - f(z)| > 4$. Hence $f(x)$ does not approach any limiting value as x approaches 1. That is, given any number L and any $\epsilon < 2$, there does not exist a corresponding δ as required by the definition of limit. △

The definition of limit states the theoretical conditions which must be satisfied for a given number, L, to be the $\lim\limits_{x \to c} f(x)$. However, the definition gives no clue to finding the number L. Thus we turn to our CAS for help in determining the number L, if it exists.

Example 3

We will find the limit, if it exists, of $f(x) = \dfrac{x^3 \sin(x) - 7x}{x \cos(x)}$ as x approaches 0. If the limit does exist, we will determine how close x must be to 0 for $f(x)$ to be within 0.001 of the limit. First we define the *function f* and take the limit:

```
> f := proc(x) (x^3*sin(x)-7*x)/(x*cos(x)) end;
> limit(f(x),x=0);
```

$$-7$$

Now we want to determine how close x must be to 0 for $f(x)$ to be within 0.001 of −7. We want

$$|f(x) - (-7)| < 0.001$$

If we use the plotting function of our CAS, starting with $-0.1 \le x \le 0.1$,

```
> plot(abs(f(x) + 7),x=-0.1..0.1);
```

we see that, as we decrease the x interval to about $-0.01 \le x \le 0.01$, the value of $|f(x) + 7|$ becomes within 0.001 of zero.

We may notice that as x becomes small, the graph of $y = |f(x) + 7|$ begins to look like a parabola. Why is this? See the section on Taylor's Theorem!

Another (graphical) method for determining how close x must be to 0 for $f(x)$ to be within 0.001 of -7 is to convert the absolute value, inequality problem ($|f(x) - (-7)| < 0.001$) into the problem of finding zeros of a function. For example, define the function g by $g(x) = |f(x) - (-7)| - 0.001$. The problem is now to determine for what values of x near 0 is $g(x) < 0$? Geometrically this is the problem of determining for what values of x near 0 does the graph of g lie under the x-axis? The answer is found by plotting g and approximating the x intercepts (i.e., the zeros of g) near 0. If your CAS has the capability of digitizing a point, the approximations are easy to obtain. Otherwise, several applications of zooming may be necessary in order to obtain a suitable approximation. $\triangle$

Exercises

In Exercises 1 through 10 evaluate the following limits or explain why a limit does not exist. If the function has a limit, determine how close x must be to the value it is approaching in order for $f(x)$ to be within 0.001 of the value it is approaching.

1. $\lim_{x \to 2} (x - 2)(x^4 + 3x^2 - 7)$

2. $\lim_{x \to 1} (x^{-1})(\cos(x) - 7\sin(x))$

3. $\lim_{x \to 0} x \left(\dfrac{\cos(x) - 3x}{\sin(x)} \right)$

4. $\lim_{x \to 4} (x - 4)\dfrac{3x - 7}{(x - 4)^2}$

5. $\lim_{x \to \pi} 6\sin(x)(\tan(x) - 3x^2 + 13)$

6. $\lim_{x \to 0} x \sin\left(\dfrac{1}{x}\right)$

7. $\lim_{x \to 0} \dfrac{\sin(x)}{x}$

8. $\lim_{x \to 3} f(x)$ where $f(x) = \sin(x)$ for $x \le 0$ and $f(x) = \cos(x)$ for $x > 0$

9. $\lim\limits_{x \to -2} f(x)$ where $f(x) = x^2 - 2$ for $x < -2$ and $f(x) = \frac{7}{x+3}$ for $x > -2$

10. $\lim\limits_{x \to 5} |x - 5|$

11. If $\lim\limits_{x \to c} f(x)$ does not exist, must f be unbounded in every interval centered at $x = c$? Illustrate your answer with examples.

12. Give a graphical argument showing that

$$\lim_{x \to 0} \sin\left(\frac{1}{x}\right)$$

does not exist. Hint: Note that the plot of $\sin(\frac{1}{x})$ oscillates between -1 and 1 over *every* open interval that contains zero.

13. Exercises 1 through 6 were in the form $\lim\limits_{x \to c} (g(x)h(x))$. On the basis of these exercises or others that you make up, conjecture and prove a theorem about the existence of a limit of a product. Consider the cases where both factors have limits, where one factor has a limit and the other does not, where neither factor has a limit.

14. $\lim\limits_{x \to c^-} f(x)$ denotes a "left-hand" limit (x approaches c through values that are less than c). In a similar manner, $\lim\limits_{x \to c^+} f(x)$ denotes a "right-hand" limit. Based on the results of Exercises 8 through 11 and similar ones that you make up, conjecture and prove a theorem relating the three limits: right-hand limit, left-hand limit, and limit.

15. This is an exploratory exercise.

   (a) Evaluate $\lim\limits_{x \to 0} \frac{\sin(x)}{x}$.

   (b) Make a conjecture [based on the results of part (a)] how $\sin(x)$ behaves in comparison to x for values of x near 0.

   (c) Draw a multigraph of $f(x) = \sin(x)$ and $g(x) = x$ (i.e., graph both functions in the same graph box). Is your confidence in your conjecture strengthened or weakened by the graphs you drew? Why?

   (d) Repeat part (c) restricting the domain to $[-.25, .25]$.

   (e) Conjecture [based on the results of parts (a) - (c)] the value of $\lim\limits_{x \to 0} \frac{\sin(\sin(x))}{x^k}$ for $k = \frac{1}{2}, 1,$ and 2. Check your conjecture by using a CAS to evaluate the limits.

   (f) Give an argument [based on your responses to (a) - (e)] supporting or rejecting the following statement:

   $$\lim_{x \to 0} \frac{\sin(\sin(x))}{x^k} \quad \text{exists and is nonzero only when } k = 1$$

16. (M. Henle) Consider the quadratic polynomial $p = kx^2 + (k+1)x - (k+2)$, where k is a constant. Examine p for several values of k: graph these polynomials and note that they have real roots. Let s be the largest of the roots of p. Find the limit of s as k approaches infinity.

17. The purpose of this exercise is to state and then verify a conjecture concerning the relationship between values for m and n and the existence or non-existence of $\lim\limits_{x \to 0} \dfrac{1 - \cos(x^m)}{x^n}$.

    (a) Develop data. Evaluate the limit (or show that the limit does not exist) for several pairs of values of m and n. (Keep in mind that you are trying to discover a pattern.)

    (b) Formulate a conjecture concerning the relationship between m and n and the existence or non-existence of $\lim\limits_{x \to 0} \dfrac{1 - \cos(x^m)}{x^n}$.

    (c) Verify your conjecture.

18. Follow the instructions in Exercise 17 for $\lim\limits_{x \to 0} \dfrac{(1 - \sin(x^m))}{x^n}$.

19. Investigate (i.e., follow the instructions in Exercise 17) $\lim\limits_{x \to 0} \dfrac{\sin(mx)}{\sin(nx)}$ in terms of the relationship between m and n.

Discussion: Part 2

The basic limit question expressed in approximation language is:

> Given the function f, what number (if any) does f approximate when x approximates a?

Expressing approximations in terms of sequences (of approximations) provides a sequence approach to limits that is an alternative to the ϵ, δ approach described in Part 1. The basic idea is to transform a limit problem into a sequence convergence problem. We will illustrate the idea with two examples before stating a formal definition. First, a word about notation. Since CAS syntax does not allow for subscripts, we will denote sequences such as f_n by fn, z_n by zn, etc. when expressing them in CAS syntax. We will rely on the context to make it clear whether an expression such as zn refers to a sequence or a product.

Example 4

We will determine $\lim\limits_{x \to 2} f(x)$ where $f(x) = \dfrac{x^2 + x}{x + 1}$ by

Step 1.   Define the sequence $s_n = \frac{1}{n} + 2$.

Step 2. Compose f with s_n forming a new sequence $f_n = f(x_n)$.

Step 3. Determine if f_n converges.

We begin by defining f, s_n, and f_n.

```
> f:= proc(x) (x^2 + x)/(x+1) end;
> s:= proc(n) 1/n + 2 end;
```

and

```
> f(s(n));
```

$$\frac{(1/n + 2)^2 + 1/n + 2}{1/n + 3}$$

Simplifying this expression and labeling it fn

```
> fn:= simplify(");
```

$$fn := \frac{1 + 2n}{n}$$

(Recall that in Maple, " refers to the previous output.) Clearly fn converges to 2. Does this mean that we have shown that the $\lim_{x \to 2} f(x)$ is 2? No, not quite. All we have shown is that $f(x)$ approximates 2 when x approximates 2 by the particular sequence $s_n = \frac{1}{n} + 2$. We have not shown what happens when x approximates 2 by some other sequence converging to 2. Before concluding that the $\lim_{x \to 2} f(x)$ is 2, we need to consider x approximating 2 by all possible sequences converging to 2, but never equal to 2 (i.e., no element of a sequence is equal to 2). We will do this by changing step 1 above to read:

Step 1. Define the sequence $s_n = z_n + 2$ where z_n converges to 0 and $z_n \neq 0$ for any n. That is,

```
> s:= proc(n) zn + 2 end;
```

Now implementing step 3 with this new definition of s_n

```
> fn:= simplify(f(s(n)));
```

$$fn := zn + 2$$

Since z_n converges to 0, fn converges to 2 and hence $\lim_{x \to 2} f(x) = 2$.   △

Example 5

We will find the $\lim\limits_{x \to 1} f(x)$ where $f(x) = \dfrac{x^2 - 1}{x - 1}$. (Note that f is not defined at $x = 1$.)

Implementing the three steps described in Example 4, we begin by defining $s_n = z_n + 1$ where z_n converges to 0 and $z_n \neq 0$ for any n. We then compose f with s_n, simplify, and label the result by fn.

```
> fn:= simplify(f(s(n)));
```

$$fn := zn + 2$$

Since z_n converges to 0, fn converges to 2, and so $\lim\limits_{x \to 1} f(x) = 2$. $\triangle$

Definition 2.10.2 *Let $f : \Re \to \Re$. Let a be a point such that $a = \lim\limits_{n \to \infty} x_n$ for some sequence $\{x_n\}$ in the domain of f, $x_n \neq a$. Then, the $\lim\limits_{x \to a} f(x) = L$, if for all sequences $\{s_n\}$ in the domain of f converging to a, $s_n \neq a$,*

$$\lim_{n \to \infty} f(s_n) = L$$

We make 5 observations about this definition:

1. The definition does not require that the function be defined at $x = a$ (see Example 5).

2. The limit is not required to be $f(a)$, even if f is defined at $x = a$.

3. The sequence $\{s_n\}$, $s_n \to a$, has none of its values equal to a.

4. The limit must be L *for all* sequences $\{s_n\}$ in the domain of f converging to a, not just for some.

5. The condition that some sequence $\{x_n\}$ in the domain of f, $x_n \neq a$, converges to a is needed to avoid having to worry about isolated points.

Exercises

20. Evaluate the limit of $f(x) = \dfrac{(x^2 - 4)\sin(x)}{x - 2}$ as $x \to 2$, provided the limit exists.

21. Evaluate the limit of $f(x) = \dfrac{x^2 - 2}{2x + 1} - \dfrac{x^2 - 4}{x - 2}$ as $x \to 2$, provided the limit exists.

22. Let $f(x) = \sin(\dfrac{1}{x})$.

(a) Compute $\lim\limits_{n\to\infty} f(s_n)$ for $s_n = \dfrac{2}{(1+4n)\pi}$.

If your CAS cannot evaluate this limit, conjecture the result by evaluating the limit for several large values of n. For example, the following `for` loop (see section 1.4.1 Control Structures) evaluates the limit as n goes from 9,000 to 10,000 in jumps of 100.

```
> for n from 90 to 100 do
≫ print(limit(sin((1+4*k)*Pi/2),k=100*n));
≫ od:
```

Now experiment with other values of n, make a conjecture, and then prove your conjecture with an analytical argument.

(b) Compute $\lim\limits_{n\to\infty} f(s_n)$ for $s_n = \dfrac{2}{(3+4n)\pi}$.

(c) What can you conclude from parts (a) and (b)?

2.11   Bisection

Prerequisites

- The concept of a continuous function

- The Intermediate Value Theorem

- The Getting Started section on the `For` loop is recommended

- The section on Roots (Zeros) of Polynomials is recommended

Discussion

This section will examine the use of the Bisection Algorithm for solving equations. In mathematics we often need to solve an equation of the form $g(x) = b$, where $g : \Re \longrightarrow \Re$. If we define $f(x) = g(x) - b$, then

$$g(x) = b \ \text{ if and only if } \ f(x) = 0$$

Thus solving $g(x) = b$ is equivalent to solving $f(x) = 0$. A solution can be found analytically only in a few cases. For example, if $f(x) = ax^2 + bx + c$ then we can solve $f(x) = 0$, or $ax^2 + bx + c = 0$, by the quadratic formula, obtaining

$$x = \frac{-b \pm \sqrt{b^2 - 4ac}}{2a}$$

We could also apply our CAS

```
> solve(a*x^2+b*x+c=0,x);
```

and obtain the same results. However, such an explicit formula is quite rare; usually it is necessary to find an approximation to the desired solution.

The problem we will consider is "existence." How can we guarantee the existence of a solution to the equation $f(x) = 0$? (Such a solution is called a *zero* of the function or a *root* of the equation.) Suppose f is defined and continuous on the interval $[a, b]$. Then by the Intermediate Value Theorem, f must take on all the values between $f(a)$ and $f(b)$. In particular, if $f(a)$ and $f(b)$ have opposite signs, then f must take on the value 0 at some point between a and b.

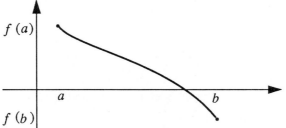

If we know that a solution must exist when f is continuous on $[a, b]$ and has opposite signs at a and b, we can use the *Bisection Algorithm* to find a solution. In outline, the algorithm proceeds as follows. (We will give more details later.) Let $ACC > 0$ be the desired accuracy in finding a solution; that is, if p is the approximate solution and x is the exact solution, then $|x - p| \leq ACC$. Let m_1 be the midpoint of the interval $[a, b]$. We consider m_1 to be the first approximation to a solution. If $f(m_1) = 0$, then m_1 is a solution. Otherwise, $f(m_1)$ has the opposite sign of either $f(a)$ or $f(b)$. Suppose $f(a)$ and $f(m_1)$ have opposite signs. Then we can repeat the above argument with the new interval $[a, m_1]$. In this way we obtain $\{m_n : n = 1, 2, ...\}$, a sequence of approximations to the answer.

Error Analysis

Each time we bisect the interval, the uncertainty of the position of the exact solution, which lies in the interval, is cut in half. Thus the maximum error for the first approximation m_1 is $\frac{b - a}{2}$.

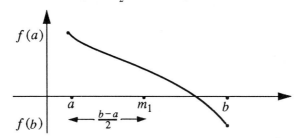

The maximum error for the second approximation m_2 is $\frac{b - a}{4}$, and in general the maximum error for the nth approximation m_n is $\frac{b - a}{2^n}$. Thus to have an error of at most ACC, we need

$$\frac{(b-a)}{2^n} \le ACC$$

or

$$n \ge \log_2(\frac{b-a}{ACC}).$$

For example, if the interval is [2,5] and $ACC = 10^{-3}$ then the number of bisections needed is at most

$$\log_2(3000) \approx 11.5$$

or 12 bisections.

Example 1. Solutions to $x^5 + 4x^3 + 2x = 8$

Suppose we wish to find a solution to

$$x^5 + 4x^3 + 2x = 8$$

with an error of at most $ACC = 0.001$. This is equivalent to approximating a zero of

$$f(x) = x^5 + 4x^3 + 2x - 8$$

with accuracy 0.001

First we define the desired function using a CAS:

```
>  f:= proc(x) x∧5+4*x∧3+2*x-8 end;
```

Since f is a polynomial, it is continuous on all of $\Re$ and we can apply the Intermediate Value Theorem if we can find points a and b with $f(a)$ and $f(b)$ of opposite signs.

There are two methods of locating an interval where f has opposite signs at the endpoints: analytical and graphical. (We used the graphical method in the section on Roots (Zeros) of Polynomials). We will discuss the analytic method first, looking at graphical methods in the next example. We note that $f(0) = -8$. Since the coefficient of the highest degree term is positive, f must be positive for large x. We can now try some values of x:

x	$f(x)$
0	-8
1	-1
2	60

(We may use a CAS and a `for` command to construct the table.) From the table we see that f has a zero in the interval [1,2]. From our error analysis above, we know that at most $\log_2(1000) \approx 10$ bisections will be required.

Having found an interval where f has opposite signs at the endpoints, we apply the Bisection Algorithm. To do this with the aid of a CAS we will define an auxiliary function to compute the midpoint:

```
> mid:= proc(a,b)(a+b)/2 end;
```

Now, starting with the initial endpoints $a = 1$ and $b = 2$ we will

   (1) use "mid" to find the midpoint, m;

   (2) evaluate f at the midpoint;

   (3) determine the new a or b;

   (4) if $b - m \leq ACC$, then we quit; otherwise we continue with step (1).

Here are the results of applying the algorithm (with breaks between applications of the algorithm and the user's input on the left):

input	output
`> m:= 1.5;`	$m := 1.500000000$
`> f(m);`	16.09375
`> b:= 1.5;`	$b := 1.5$
`> m:= mid(a,b);`	$m := 1.250000000$
`> f(m);`	5.36425781
`> b:= m;`	$b := 1.250000000$
`> m:= mid(a,b);`	$m := 1.125000000$
`> f(m);`	1.747344971
`> b:= m;`	$b := 1.125000000$
`> m:= mid(a,b);`	$m := 1.062500000$
`> f(m);`	.276932718
`> b:= m;`	$b := 1.062500000$
`> m:= mid(a,b);`	$m := 1.031250000$
`> f(m);`	$-.384333576$
`> a:= m;`	$a := 1.031250000$
`> m:= mid(a,b);`	$m := 1.046875000$
`> f(m);`	$-.059568832$
`> a:= m;`	$a := 1.046875000$
`> m:= mid(a,b);`	$m := 1.054687500$
`> f(m);`	.107193382
`> b:= m;`	$b := 1.054687500$
`> m:= mid(a,b);`	$m := 1.050781250$
`> f(m);`	.023442832
`> b:= m;`	$b := 1.050781250$

```
> m:= mid(a,b);     m := 1.048828125
> f(m);             -.018155022
> a:= m;            a := 1.048828125


> m:= mid(a,b);     m := 1.049804688
> (m);              .002620869
> b-m;              .000976563
```

Thus, a zero of f in [1,2] is 1.0498 with an error of at most 0.001. $\triangle$

The repetitions in the last example can be left to the computer by using the **while** command. (See the section on Getting Started.) After a, b, f, and mid have been defined as above, we can give the commands:

```
> # Set the accuracy.
> ACC:= 0.001;
> while b - a > 2*ACC do        # while interval is too wide...
≫       m:= mid(a,b);           # compute midpoint and ...
≫       if f(m)*f(a) =< 0 then  # compute new endpoints.
≫               b := m
≫       else
≫               a := m
≫       fi;
≫ od;
```

Many systems will have a "bisection" or other numerical zero finding algorithm available. If yours does not, the preceding can be rewritten as a user-defined function, "bisect."

```
> bisect:= proc(f,left,right,acc)
≫       # Bisection Algorithm.
≫       # Input:  f = function whose zero is to be found.
≫       # left = left end point of interval of search.
≫       # right = right end point of interval of search.
≫       # acc = accuracy of approximate zero.
≫       # Variables local to the procedure.
≫       local mid,a,b;
≫       a:= left;
≫       b:= right;
≫       while b - a > 2*acc do
≫               mid:= (a+b)/2;
≫               if evalf(f(a)*f(mid)) < 0 then
≫               # zero is in [a,mid], change b.
≫                       b:= mid
≫               else
≫               # zero is in [mid,b], change a.
≫                       a:= mid
≫               fi;
```

```
≫          od;
≫          RETURN(mid);
≫ end:
```

Example 2. All Solutions to an Equation

In the previous example we were satisfied with finding *one* solution to an equation. We will now consider the problem of finding *all* solutions to an equation (in
an interval (a, b)). We will reduce this problem to the problem of the previous
example by decomposing (a, b) into subintervals in which f has a single zero
(root). Then the Bisection Algorithm can be applied to each subinterval. (The
section on Roots (Zeros) of Polynomials also considers this problem.)

The easiest method to find subintervals over which f has a single root is to
graph the function. Suppose we wish to find all roots of $f(x) = x^5 - 3x^4 + 2x^3 + 1$
on the interval $(-\infty, +\infty) = \Re$. We define the function f:

> `f:= proc(x) x∧5 - 3*x∧4 + 2*x∧3 + 1 end;`

Since f is a polynomial of odd degree, the value will be negative when x is very
negative and the value will be positive when x is very positive. The shape of the
curve will be:

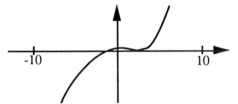

when plotted over the default interval, $[-10, 10]$. However since p is negative for
$x < -1$ and positive for $x > 1$ (Why?), we replot over $[-1, 2]$. We note that
since p is a fifth degree polynomial it can have at most 5 real zeros.

The polynomial can have at most 5 real zeros since the degree is 5. We graph
f over successively smaller intervals looking for zeros. The graph is similar to
the above until we consider the interval $[-1, 2]$:

> `plot(f(x),x=-1..2);`

returns

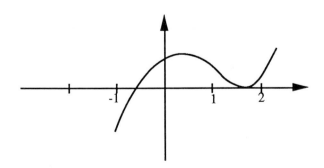

There is a zero near -0.7 and perhaps several zeros between 1.5 and 2. Plotting over the interval $[1.5, 2]$ gives:

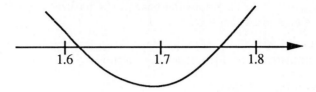

We can now apply the Bisection Algorithm of the previous example to find the zeros. Using the user-defined bisection procedure on f with an accuracy of 0.001 gives:

input	output
bisect(f,-0.8,-0.6,0.001);	-0.6180
bisect(f,1.6,1.7,0.001);	1.6183
bisect(f,1.7,1.8,0.001);	1.7550

In this way we can find all solutions of an equation.   △

There are two situations that one needs to be alert to when graphically searching for zeros of a function. One is the existence of a zero of even multiplicity (the graph touches, but does not cross the x-axis) and the other is a zero that lies outside of the graphing interval. The situation of a zero of even multiplicity can usually be handled by zooming. However, determining all of the zeros is often a more difficult situation and usually requires using analytical methods as well as graphical. For example, determining the number of zeros that exist or determining how a function behaves (positive or negative) for x outside of the graphing interval.

Exercises

In the following exercises, you should provide a description of the problem, the method of solution, and the results.

1. Find a 6th root of 100 with an accuracy of 0.001.

2. Find a solution of $x^5 + x^2 + 1 = 0$ on $[-2, -1]$ with an accuracy of 0.001.

3. Find all solutions to $\sin(x^2) + 1 = x^2$ with an accuracy of 0.001.

4. Find all solutions to $\sin(x^2) = x^2$ with an accuracy of 0.001.

5. Find all solutions to $x^5 - 2x^2 + 3x + 1 = 0$ with an accuracy of 0.001.

6. Find all solutions to $x^6 - 5x^4 + 2x - 1 = 0$ with an accuracy of 0.001.

7. The objective of this exercise is to find a numerical approximation, accurate to four decimal places (i.e., $error < 0.00005$), to the base of the natural logarithm.

Recall that the relation between logarithms and exponents is given by $y = \log_a(x)$ if and only if $x = a^y$. Thus, in particular, if a is the base, then $\log_a(a) = 1$. Therefore the base of the natural logarithm, $\log(x)$, is the value of x for which $\log(x) = 1$.

Approximate with four decimal place accuracy the base of the natural logarithm using each of the following methods.

(a) Bisection algorithm.

(b) Graphical: superimpose the graph of $y = \log(x)$ on the graph of $y = 1$ and then "zoom in" on the intersection point.

(c) Use your CAS's solve command.

2.12   Growth of Functions

Prerequisites

- Polynomial functions for Part 1

- Exponential and logarithmic functions for Part 2

- The section on the Bisection Algorithm for Part 2

- The section on Elementary Graphing

Discussion: Part 1

In this section we will investigate the growth of functions. We will compare the behavior of a function with the behavior of standard functions such as $f(x) = x$, $f(x) = x^2$, and $f(x) = x^3$. You may find it useful to graph these (using your CAS) as the discussion proceeds. In the second part, we will consider the exponential and logarithmic functions.

Consider the function $y = f(x) = 3x$. If the value of x is doubled, the value of y is doubled. More generally, if the value of x is multiplied by $c > 0$, then the value of y is multiplied by c. When this behavior occurs we say that y is *proportional* to x, and may write $y \propto x$. (Formally, $y \propto x$ means that there is a constant C such that $y = Cx$.) Which of our basic functions has similar behavior? Clearly if $y = f(x) = x$, then y is proportional to x, so the behavior of $y = 3x$ is similar to $y = x$: y grows proportionally to x.

Now we will modify the function by adding a constant. Consider the function $y = f(x) = 3x + 1$. Now y is no longer proportional to x, since if we double x, from 1 to 2, the value of y changes from 4 to 7; if we triple x, from 2 to 6, the value of y changes from 7 to 19. If y were proportional to x, these latter values would have been 8 and 21 respectively. We see that, although y is not proportional to x, the relation is still similar. In fact, the larger the values, the more similar the functions are. If $x = 1000$, then $f(1000) = 3 \times 1000 + 1 = 3001$

and doubling x gives $f(2000) = 6001$. If the change were proportional, the new value would be $3 \times 2000 = 6000$. We can compare these numbers by considering their relative difference. By the *relative difference* of two numbers x and y we mean

$$rel(x, y) = \frac{|x - y|}{\min(x, y)}$$

For the numbers 6000 and 6001, the relative difference is

$$rel(6000, 6001) = \frac{6001 - 6000}{6000} = \frac{1}{6000}$$

Thus $rel(3x + 1, 3x) = \dfrac{1}{3x}$ and, clearly, 1 is negligible compared with $3x$ for large values of x.

We can define the function rel on the CAS for general use.

```
> rel:= proc(x,y) abs(x-y)/min(x,y) end;
```

Two functions which have a small relative difference have similar growth rates for large values of x. There is a notation for describing such functions, the *Big O* notation (read "big oh"). The O is an abbreviation for *Order*. The idea is that a function f is Big O of a function g, written $f = O(g)$ if f grows "no faster" than g for large values. The precise definition is given for positive functions.

Definition 2.12.1 *If functions f and g are non-negative, then f is Big O of g, written $f = O(g)$, if there is a constant $C > 0$ such that for large x, $f(x) \le Cg(x)$.*

Note the "$\le$" in the definition; thus $x = O(x)$ (since $x \le x$ for all x) $x = O(x^2)$ (since $x \le x^2$ for $x \ge 1$), $x = O(x^3)$, etc.

Two functions f and g have a *similar* growth rate if $f = O(g)$ and $g = O(f)$. If this holds then we write $f = \Theta(g)$ or, equivalently, $g = \Theta(f)$. Note that if $f = Cg$, $C > 0$, then $f = \Theta(g)$. In the case of $f(x) = 3x + 1$, we see that if we set $C = 4$ and $g(x) = x$ then $f(x) = 3x + 1 \le 4x = 4g(x)$ when $x \ge 1$. Thus we can write $3x + 1 = O(x)$. We also have $x = O(3x + 1)$, since $x \le 3x + 1$, and so $3x + 1 = \Theta(x)$. Similarly, if $x \ge 10$ then $5x + 10 \le 6x$, so $5x + 10 = O(x)$, and, in fact, $5x + 10 = \Theta(x)$.

In general practice, the Big O notation is usually used to describe the growth rate of a function, f, by comparing it to the growth rate of a standard, function, g. Thus we write $f(x) = \dfrac{7x^3 - 32x^2 + 13}{4x - 3} = O(x^2)$ and say that f grows no faster than g for large values of x.

Exercises

1. Examine the graphs of $y = x$, $y = 3x$, $y = 3x + 1$, and $y = 5x + 10$. What can you say about their growth rate?

We will now consider functions with a quadratic (squared) term. The standard quadratic (second degree) function is $g(x) = x^2$. The function $f(x) = 3x^2$ is proportional to g, so $f = O(g)$. Similarly, $g = O(f)$, so $f = \Theta(g)$. The function $f(x) = 3x^2 - 2x + 10$ is not proportional to g, but the behavior is similar. When x is large, the highest degree term, $3x^2$, is dominant, and so f is similar to $3g(x)$, i.e., $f = \Theta(g)$. How large must x be for the relative difference between $f(x)$ and the dominant term $3g(x)$ to be less than 10^{-4}? That is, what is the smallest x such that

$$rel(f(x), 3g(x)) = \frac{|f(x) - 3g(x)|}{\min(f(x), 3g(x))} < 10^{-4}?$$

First we can estimate this value analytically. When x is large, $|f(x) - 3g(x)| = 3g(x) - f(x) = 2x - 10$ and $\min(f(x), 3g(x)) = f(x) = 3x^2 - 2x + 10$. Thus

$$rel(f(x), 3g(x)) = \frac{3g(x) - f(x)}{f(x)} = \frac{2x - 10}{3x^2 - 2x + 10}$$

When x is large, the numerator is dominated by $2x$ and the denominator is dominated by $3x^2$. (For example, when $x = 1000$, the numerator is $2,000 - 10$, almost the same as the first term alone, and the denominator is $3,000,000 - 2,000 + 10$, again relatively close to the first term alone.) Thus, when x is large, we want

$$rel(f(x), 3g(x)) \approx \frac{2x - 10}{3x^2} \leq \frac{2x}{3x^2} = \frac{2}{3x} < 10^{-4}$$

This occurs when $x > 6666$.

Now, if we want a more precise answer we can use our CAS to print out the values using a **for** loop, varying x from 6660 to 6670. (We use **evalf** to ensure a numerical response rather than a symbolic response from the CAS.)

```
> # First we define f, and g.
> f:= proc(x) 3*x^2 - 2*x + 10 end;
> g:= proc(x) x^2 end;
> # Now we compute and print rel.
> for x from 6660 to 6670 do
≫         print(x,f(x),3*g(x),evalf(rel(f(x),3*g(x))));
≫ od;
```

Some of the output is

6660,	133053490,	133066800,	.0001000349559
6661,	133093451,	133106763,	.0001000199476
6662,	133133418,	133146732,	.0001000049439
6663,	133173391,	133186707,	.00009998994469
6664,	133213370,	133226688,	.00009997494996

We find that the relative difference drops below 10^{-4} between 6662 and 6663.

Exercises

2. Perform an analysis similar to the above for the function $f(x) = 4x^3 - 3x^2 + 2x - 10$:

   (a) What is the dominant term of f?

   (b) Construct a table of x, $f(x)$, the dominant term, and the relative difference so as to experimentally determine how large x must be for the relative difference to be less than 10^{-4}.

   (c) Analytically estimate how large x must be for the relative difference to be less than 10^{-4}.

Discussion: Part 2

In the previous section we compared functions to polynomials of the form $f(x) = x^n$ for $n > 0$. There are functions which grow faster than any positive power (e.g., the exponential functions) and functions which grow slower than any positive power (e.g., the logarithmic functions). For example,

```
> limit(x∧(10)/2∧x,x=infinity);
```

returns 0. Thus for any positive integers $n > 0$ and $b > 1$, $x^n = O(\exp_b(x))$ (where $\exp_b(x) = b^x$) and $\log(x) = O(x^n)$. In terms of the graphs of the functions, $\exp_b(x) = b^x$ will eventually exceed any polynomial.

Exercises

Your CAS may help with the calculations in the following exercises. In Exercises 3 through 5, the term *relative error* means the relative difference between the exact value and the approximation.

3. Find, with a relative error of 10^{-3}, the value of x where $2^x > 3x^5$.

4. Find, with a relative error of 10^{-3}, the value of $x \geq 1$ where

$$10 \log(x) + 5 \leq x$$

5. Find, with a relative error of 10^{-3}, the value of $x \geq 1$ where

$$10 \log(x) + 5 \leq \sqrt{x}$$

6. Which grows faster, $f(x) = x^r$ where r is a positive number, or $g(x) = \log(x)$? Try your CAS on several values of r (especially small values of r).

7. Show how the fact that exp grows faster than any polynomial implies that log grows more slowly than any polynomial.

8. Which grows faster, n^n or 2^n? Try using your CAS. If it fails, then try factoring each expression.

9. Which grows faster, $n!$ or n^n? Hint: See the previous exercise.

10. Which grows faster, $n!$ or 2^n? Hint: See the previous two exercises.

11. Rank the following functions of n in order of growth, with the fastest first. Use the discussion and exercises of this section. n, $\sin(n)$, $\log(n)$, 3^n, $n!$, n^4, and n^n.

Chapter 3

Differentiation

3.1   The Derivative

Prerequisites

- The section on Limits

Discussion: Secant Lines

In this section we will explore the meaning of the derivative of a function f, concentrating on geometric interpretations. The derivative of a function is the *instantaneous rate of change* of the output of the function with respect to the input of the function. If the derivative of a function exists at every point in the domain, the function is said to be differentiable. The process of finding the derivative of a function is called differentiation. Our development will be to approximate the instantaneous rate of change of a function f at $x = a$ by the *average rate of change* of f over an interval with a as an endpoint. We will then improve the approximation by decreasing the size of the interval.

We start with the the average rate of change of f over the interval $[a, a + h]$

$$\text{av. rate of change} = \frac{fa + h) - f(a)}{(a + h) - a} = \frac{f(a + h) - f(a)}{h}$$

The expression $\dfrac{f(a + h) - f(a)}{h}$ is usually referred to as the *difference quotient*. What is the geometric meaning of the difference quotient? Suppose f has the graph on the left:

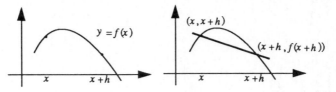

The evaluations of f at the two values a and $a + h$ determine two points $(a, f(a))$ and $(a + h, f(a + h))$ on the graph of f. These two points in turn determine a straight line, as shown above on the right. This line, intersecting the graph in (at least) two points, is called a *secant* line. The slope of this secant line is

$$\frac{\text{change in } y}{\text{change in } x} = \frac{f(a + h) - f(a)}{(a + h) - a} = \frac{f(a + h) - f(a)}{h}$$

Thus the difference quotient is the slope of the secant line determined by a and $a + h$. The equation of the curve is $y = f(x)$. What is the equation of the secant line? The equation of a straight line is

$$y = (\text{slope})x + b$$

where the slope is the difference quotient, $\dfrac{f(a + h) - f(a)}{h}$. The value of the constant b is determined by substituting into the above equation the coordinates of a point on the line. For example, if $(a, f(a))$ is a point on the line,

$$b = f(a) - \frac{f(a + h) - f(a)}{h} a$$

Thus the equation of the secant line is

$$\begin{aligned} y &= \frac{f(a + h) - f(a)}{h} x + (-a)\frac{f(a + h) - f(a)}{h} + f(a) \\ &= \frac{f(a + h) - f(a)}{h}(x - a) + f(a) \end{aligned}$$

We want to plot several of these secant lines, so we define a function with variables x, a, and h:

```
> S:= proc(x,a,h) (f(a+h) - f(a))/h * (x-a) + f(a) end;
```

Clearly, if a is fixed and h is made small, then $x + h$ gets close to x, the two points get close together and the secant lines get close together. Let's try this with a simple function.

Example 1. Secant Lines for $f(x) = x^2$

We can use the above definition of S for the secant line, but first we must define our function f:

```
> f := proc(x) x^2 end;
```

We can now plot, on the same axes, the graphs of f and of several secant lines. We will fix $a = 1$ and let h take several values:

```
> plot({f(x),S(x,1,1),S(x,1,0.5)},x=0..3);
```

This returns

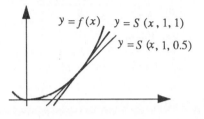

Notice that as h gets smaller, the secant lines approach a limiting position. This limiting position is called the line tangent to the graph at $a = 1$.   △

Intuitively a *tangent line* to a curve is a line that just touches the curve and has the same "direction" as the curve. Formally, the tangent line to the graph of $y = f(x)$ at the point $x = a$ is the straight line passing through the point $(a, f(a))$ with slope $\lim\limits_{h \to 0} \dfrac{f(a + h) - f(a)}{h}$. If this limit does not exist, then there is no tangent line to the graph at the point $x = a$.

Exercises

In the following exercises, plot the function and secant lines. Choose an interval symmetric about the point a.

1. Let $f(x) = x^3$, $a = -1$, and h have the values 1, 0.5, and 0.1. Plot f and the secant lines on the same set of axes. What do you think the slope and the equation of the tangent line are?

2. Let $f(x) = \sin(x)$, $a = 0$, and h have the values 1, 0.5, and 0.1. Plot f and the secant lines on the same set of axes. What do you think the slope and the equation of the tangent line are?

3. Let $f(x) = |x|$, $a = 0$, and h have the values 1, 0.5, and 0.1. Plot f and the secant lines on the same set of axes. (Hint: For a CAS, $|x|$ is written as $abs(x)$.) What do you think the slope and the equation of the tangent line are?

4. Let $f(x) = |x|$, $a = 0$, and h have the values -1, -0.5, and -0.1. Plot f and the secant lines on the same set of axes. What do you think the slope and the equation of the tangent line are?

Discussion: The Derivative

If the secant lines *always* approach a common line as h approaches zero, then f has a derivative at a and the value of the derivative, written $f'(a)$, is the common limit of the slopes of the secant lines. (Exercises 3 and 4 above exhibit a case where no such *common* limit exists.) Thus we can calculate the derivative as

$$f'(a) = \lim_{h \to 0} \frac{f(a + h) - f(a)}{h}$$

The equation of the tangent line becomes

$$y = f'(a)(x - a) + f(a)$$

Example 2. The Derivative of $f(x) = x^2$.

We apply our CAS to compute the limit of the difference quotient:

```
> a:= 1;
> limit((f(a+h) - f(a))/h, h=0);
```

This returns 2 as the limit. Thus the derivative of f at $a = 1$ is 2, i.e., $f'(1) = 2$. The equation of the tangent line to the graph of f at $a = 1$ is thus

$$y = f'(a)(x - a) + f(a) = 2(x - 1) + 1 = 2x - 1$$

Plotting $y = f(x)$ and $y = 2x - 1$ on the same set of axes verifies this result. $\triangle$

Exercises

5. Let $f(x) = x^3$ and $a = -1$. What does your CAS return as the limit of the difference quotient, $f'(a)$? Plot, on the same set of axes, the curves $y = f(x)$ and $y = f'(a)(x - a) + f(a)$.

6. Let $f(x) = \sin(x)$ and $a = 0$. What does your CAS return as the limit of the difference quotient, $f'(a)$? Plot, on the same set of axes, the curves $y = f(x)$ and $y = f'(a)(x - a) + f(a)$.

7. Let $f(x) = |x|$ and $a = 0$. What does your CAS return as the limit of the difference quotient?

In the previous examples and exercises it was relatively easy to tell if a derivative existed. In the following sequence of exercises, it is more difficult. In these exercises we will be concerned with functions of the form

$$f(x) = \begin{cases} x^n \sin(1/x) & \text{if } x \neq 0, \\ 0 & \text{if } x = 0 \end{cases}$$

for $n = 0, 1, 2$. Hint: To define this function, use the `if` command (see the subsection on Control Structures in the section on Getting Started).

8. Consider the above function f when $n = 0$. Plot the graph of f on the interval $[-1, 1]$. Does f have a derivative at zero? You may wish to plot f on smaller intervals to help you decide. Check your answer by using the `limit` command as in the above example.

9. Consider the above function f when $n = 1$. Plot the graph of f on the interval $[-1, 1]$. Does f have a derivative at zero? You may wish to plot f on smaller intervals to help you decide. Check your answer by using the `limit` command as in the above example.

10. Consider the above function f when $n = 2$. Plot the graph of f on the interval $[-1, 1]$. Does f have a derivative at zero? You may wish to plot f on smaller intervals to help you decide. Check your answer by using the `limit` command as in the above example.

Discussion: Approximation

One use of the derivative is to approximate a non-linear function by a linear function. A *linear function* is one whose graph is a straight line. Certainly straight lines are easier to work with than one with many curves. If we want to approximate a curve $y = f(x)$ at a point $(a, f(a))$, then the best approximating straight line is the tangent line, $y = f'(a)(x - a) + f(a)$. Thus for x near a,

$$f(x) \approx f(a) + f'(a)(x - a)$$

The function $L(x) = f(a) + f'(a)(x - a)$ is the (best) linear approximation to f for x near a. At $x = a$ we have $L(a) = f(a)$, so the approximation is exact.

Example 3. Best Linear Approximation

We will find the best linear approximation to

$$f(x) = \begin{cases} x^2 \sin(1/x) & \text{if } x \neq 0, \\ 0 & \text{if } x = 0 \end{cases}$$

for x near 1. We could use the limit to find $f'(1)$ as before, but we will use the CAS's built-in command, `diff`, and substitute $x = 1$:

```
> fp1:= subs(x=1,diff(f(x),x));
```

This returns $2\sin(1) - \cos(1)$. Thus the best linear approximation to f for x near 1 is

$$\begin{aligned} y &= f'(a)(x - a) + f(a) = (2\sin(1) - \cos(1))(x - 1) + \sin(1) \\ &\approx 1.142639664(x - 1) + .8414709848 \end{aligned}$$

where we applied `evalf` to $2\sin(1) - \cos(1)$ and $\sin(1)$ to obtain numerical approximations. △

As with any approximation, we are interested in estimating the error of our approximation. For example, how close must x be to a in order that $|f(x) - L(a)| < 0.01$? This question is answered in the section on Taylor's Theorem in a somewhat more general setting.

Exercises

11. Find the best linear approximation to $f(x) = 4x^3 + 3x^2 + 2x + 1$ for x near 0. Plot f and the linear approximation on the same set of axes.

12. Find the best linear approximation to $f(x) = 3x^3 - 2x^2 + 4x - 5$ for x near 0. Plot f and the linear approximation on the same set of axes.

13. Examine the function f and the linear approximation found in each of Exercises 11 and 12. What is the relation between the polynomial's form and the approximation's form? Can you predict the best linear approximation to $f(x) = 5x^4 - 2x^3 + 6x^2 - 4x + 7$ for x near 0 by inspection, without doing any computations? Check your answer. What is the best linear approximation to $f(x) = ax^n + bx^{n-1} + \cdots + cx + d$ for x near 0?

14. Find the best linear approximation to $f(x) = \cos(x)$ for x near $\pi/4$. Plot f and the linear approximation on the same set of axes.

15. Find the best linear approximation to $f(x) = \exp(x)$ for x near 0. Plot f and the linear approximation on the same set of axes.

3.2   Geometric Significance of the Derivative

Prerequisites

- The section on The Derivative

- The first and second derivative tests

Discussion

The numerical sign of the first or second derivative of a function provides useful information about the behavior of a function for finding extremal values as well as for predictive purposes. (For example, consider the first or second derivative tests for extremal values.) In order to gain a "feeling" for the predictive aspects of the first and second derivatives, it is helpful to think of a function f whose input is time (e.g., f may be a distance function with input time and output distance). The first derivative of f, the instantaneous rate of change, is often interpreted as *velocity* and the second derivative as *acceleration*. A positive acceleration at time t_0 predicts an increase in velocity over the time period $(t_0, t_0 + h)$ for small h, and a negative acceleration at time t_0 predicts a decrease in velocity over the time period $(t_0, t_0 + h)$ for small h. Since the the second derivative predicts something about the behavior of the first derivative which, in turn, predicts something about the behavior of the function in the "near" future, the second derivative of a function of time is often interpreted as indicating a future trend. Consider, for example, the sine function $f(x) = \sin(x)$. Does the fact that $f''(x) = -f(x)$ help explain why the sine function is periodic?

The following two examples illustrate several properties of a function that can be determined from the graph of the derivative of the function.

Example 1

We consider the following multiplot showing the graphs of $y = f(x)$ and $y = f'(x)$.

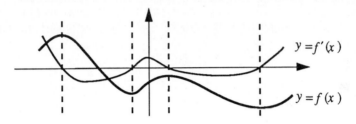

Note that

(a) The graph of $y = f'(x)$ lies below the x-axis over the intervals where f is a decreasing function (i.e., the derivative is negative).

(b) The extremal values of f occur at the x-intercepts of the graph of $y = f'(x)$ (i.e., the critical points).

(c) At the critical points where f assumes a maximum value, the graph of f' crosses the x-axis with a negative slope. However, at the critical points where f assumes a minimum value, the graph of f' crosses the x-axis with a positive slope.

$\triangle$

It would be instructive for the reader to verify points (a)-(c) by repeating Example 1 with a familiar function, say $f(x) = \sin(x)$.

Example 2

We consider the multiplot showing the graphs of $f(x) = x^3$ and $f'(x) = 3x^2$.

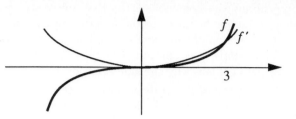

We note that the graph of f' touches the x-axis at the critical point $x = 0$, but does not cross the axis there. We also note that f does not have an extreme value at the critical point $x = 0$. Is it always true that the graph of the derivative function will touch, but not cross the x-axis at a critical point at which the function does not have an extreme value?   △

The following example illustrates that even though several graphical properties of f can be determined from the graph of f', the (vertical) location of the graph of f cannot.

Example 3

The following multiplot shows the graphs of two different functions, f and g, that have the same derivative.

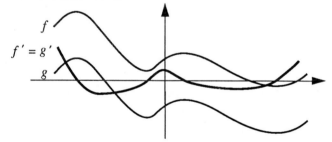

Shifting a graph vertically gives the graph of a different function, but one which has the same derivative as the original function. Thus in this example, we know from the graph of f' that f has a minimum at $x = 2$ (i.e., f' crosses the x-axis with a positive slope), but we cannot tell the value of $f(2)$ from the graph of f'.   △

There are several other questions that could be investigated concerning properties of a function that can be observed from the graph of its derivative. Is it possible to determine anything about inflection points from the graph of f'? What, if anything, do the multiplicities of the zeros of f' imply about the behavior of f? Can concavity, a property usually associated with the second derivative, be determined from the first derivative?

Exercises

1. Consider the function defined by $f(x) = (x + 5)(x + 1)^2(x - 3)^3$. Draw a multiplot of f and its first and second derivatives.

    (a) Describe the information about f that can be obtained from the graph of f'.

    (b) Describe the information about f that can be obtained from the graph of f''.

    (c) "Check out" your answers to parts (a) and (b) by analyzing the multiplot of a function and its first two derivatives for at least two other functions.

2. Consider the function defined by $f(x) = x \sin(x + 2)$. Draw a multiplot of f and its first and second derivatives.

    (a) Describe the information about f that can be obtained from the graph of f'.

    (b) Describe the information about f that can be obtained from the graph of f''.

3. Assume that a is a critical point of a differentiable function f, i.e., $f'(a) = 0$. Considering each of the possibilities: f has a minimum value at $x = a$, f has a maximum value at $x = a$, and f has neither a minimum nor a maximum value at $x = a$,

    (a) Describe the first derivative test in terms of the behavior of the graph of f' near $x = a$.

    (b) Describe the second derivative test in terms of the behavior of the graph of f' near $x = a$.

    Hint: Consider some specific functions, say $f(x) = x^2, g(x) = x^3$, and $h(x) = \sin(x)$, and specific critical points, say $x = 0$. Then generalize to an arbitrary differentiable function with a critical point at $x = a$.

4. Redo Exercise 3 on the assumption that the derivative of f does not exist at $x = a$.

5. The purpose of this exercise is to develop (or confirm) mental pictures of the shapes of graphs based on the numerical sign of the first and second derivatives of the corresponding functions.

    (a) Define three functions that have positive first derivatives, draw a multiplot of their graphs, and describe the "basic" shape of the curves.

    (b) Define three functions that have negative first derivatives, draw a multiplot of their graphs, and describe the "basic" shape of the curves.

(c) Define three functions that have positive second derivatives, draw a multiplot of their graphs, and describe the "basic" shape of the curves that is attributable to the second derivative.

(d) Define three functions that have negative second derivatives, draw a multiplot of their graphs, and describe the "basic" shape of the curves that is attributable to the second derivatives.

(e) Define two functions that have a positive first derivative and positive second derivative, draw the multiplot of their graphs, and describe the "basic" shapes.

(f) Define two functions that have positive first derivatives and negative second derivatives, draw a multiplot of their graphs, and describe the "basic" shapes.

(g) Define two functions that have negative first derivatives and positive second derivatives, draw the multiplot of their graphs, and describe the "basic" shapes.

(h) Define two functions that have negative first derivatives and negative second derivatives, draw the multiplot of their graphs, and describe the "basic" shapes.

6. For each of the following define a function on the largest domain possible that satisfies the stated properties, plot the graph of the function, and describe the significant shapes. List the functions you consider.

(a) Positive first and second derivatives, bounded first derivative.

(b) Positive first and second derivatives, bounded first and second derivatives.

(c) Positive first derivative, negative second derivative, bounded first derivative, and unbounded second derivative.

(d) Negative first derivative, positive second derivative, unbounded first derivative, and bounded second derivative.

7. Give three examples of functions that have infinitely many derivatives.

8. (Open Ended) Let $f : \Re \rightarrow \Re$ be a function defined over (a, b) having a continuous derivative with the property that $|f'(x)| < 10$. What can you say about the range of f? Describe your experiments and conjectures.

9. (Open Ended) Let $f, g : \Re \rightarrow \Re$ be functions that have continuous first derivatives defined over (a, b) and let $f'(x) < g'(x)$ for all x in (a, b). Is it true that $f(x) - f(a) < g(x) - g(a)$ for all x in (a, b)? Why?

Hint: Is $h : \Re \rightarrow \Re$ defined by $h(x) = g(x) - f(x)$ an increasing function?

10. The curves labeled a, b, and c in the following multiplot are the graphs of a function and its first two derivatives. Determine which curve represents

the function, which one represents the derivative, and which one represents the second derivative.

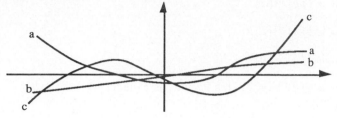

3.3   Leibniz's Rule

Prerequisites

- Differentiation of functions of one variable

- The Binomial Theorem

Discussion

Let $f^{(n)}$ denote the nth derivative of f. If f and $g : \Re \longrightarrow \Re$, then

$$(fg)^{(0)}(x) = f(x)g(x) \quad \text{(the 0th derivative)}$$

and

$$(fg)^{(1)}(x) = (fg)'(x) = f'(x)g(x) + f(x)g'(x)$$

The object is to find an expression for $(fg)^{(n)}(x)$ for arbitrary integers $n \geq 0$.

Example

We can use our CAS to compute:

```
> diff(f(x)*g(x),x);
> diff(",x);
```

which returns

$$\left(\frac{d^2}{dx^2}f(x)\right)g(x) + 2\frac{d}{dx}f(x)\frac{d}{dx}g(x) + f(x)\left(\frac{d^2}{dx^2}g(x)\right)$$

Continue this in the exercises to find a pattern.   △

Exercises

1. What is $(fg)^{(3)}(x)$?

2. What is $(fg)^{(4)}(x)$?

3. What is $(fg)^{(5)}(x)$?

4. What is $(fg)^{(n)}(x)$? This formula is called *Leibniz's Rule*.

3.4   Sketching Rational Functions

Prerequisites

- Definition of asymptotes (vertical and horizontal) and how to determine them

- The terms *zero* (root) of a function, *factor*, and *multiplicity* of a zero or a factor

- How to sketch the graph of a function like $f(x) = \dfrac{(x+2)(x+1)}{(x+3)(x-1)}$

Discussion

Someone once said "a picture is worth a thousand words." In mathematics, a good sketch may be worth a thousand computations. When a student understands the geometric interpretation of the analytical properties of a function, then the student is able to *use a sketch to guide the analysis of a function*. Global properties of a function (e.g., asymptotic behavior, numerical sign – positive or negative, and domain) are easily shown by a sketch. Likewise, local properties (e.g., intercepts, direction, concavity, and extreme points) can be approximated by a sketch. The major goal of this section is to learn how to use a CAS to help understand the geometric interpretations of the analytical properties of a function. To develop this understanding, we encourage students to approaching sketching as a multi-step process:

Step 1. Obtain a sketch, either with a CAS or by hand.
Step 2. Use the sketch to generate questions that need to be answered,
        i.e., where are the zeros, asymptotes, extremal values, etc.
Step 3. Obtained a revised sketch (with a CAS or by hand) that answers the
        questions raised in step 2.
Step 4. Repeat steps 2 and 3 until a satisfactory graph is obtained.

For rational functions, it is often easier to draw the final graph by hand having used the CAS for the initial sketch and to answer the resulting questions. We encourage students to practice developing the following "Basic Mental Approach" in curve sketching.

Basic Mental Approach

Ask (yourself):

1. What do I need to know?

2. What can I tell by inspection?

3. What are the possibilities suggested by what I know?

4. What should I do next?

Warning: Most CAS plotting routines encounter difficulties with scaling when plotting a function that is unbounded over the specified interval. Since a rational function is unbounded on an interval containing a vertical asymptote, a CAS plot may be misleading.

Sketching graphs of rational functions relying only on knowledge of the function's asymptotes, and intercepts (with their multiplicities) provide a nice introduction to sketching in the large.

Example 1

We will use the CAS to sketch the graph of

$$f(x) = \frac{x^3 - 7x}{x^3 + 3x - 3}$$

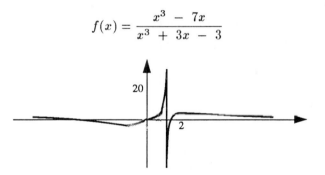

The CAS plot suggests that there is a vertical asymptote near $x = 1$ and a horizontal asymptote that is either the x-axis or near to it. The large vertical scale resulting from the fact that f is unbounded disguises the location of the zeros and the horizontal asymptote. What questions do we have?

Recall the questions in the Basic Mental Approach.

1. What do we need to know? We need to determine the x-intercepts (zeros of f) and their multiplicities.

2. What can we tell by inspection? By inspection of the defining expression for f it is clear that f has zeros of multiplicity one at $x = 0$, $+\sqrt{7}$, and $-\sqrt{7}$. (Why?)

3. What are the possibilities suggested by what we know? Since the multiplicity of each zero is odd, the curve will cross the x-axis at each of the zero points. (See the section on Roots (Zeros) of Polynomials.)

4. What should we do next? We need to determine the asymptotes, if there are any. By inspection, it is clear that for large values of x, f behaves like $\dfrac{x^3}{x^3}$. Thus $y = 1$ is a horizontal asymptote. Every vertical asymptote occurs at a zero of the denominator (but not necessarily conversely). Thus, to find the vertical asymptotes we need to find the zeros of the denominator. Since the denominator is a cubic polynomial, it will have either one or three real zeros (roots). We will plot the denominator to determine which of these possibilities exist. Since (by inspection) the denominator is positive for $x > 2$ and negative for $x < 2$, we plot the denominator over $[-2, 2]$. To do this, we first define the function and then extract the denominator and label it d.

```
> f:= proc(x) (x^3 - 7*x)/(x^3 + 3*x -3) end;
```

```
> d:= denom(f(x));
```

```
> plot(d, x=-2..2);
```

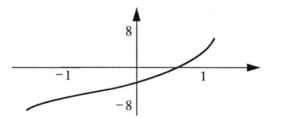

Thus there is exactly one real zero. To find a numerical approximation for this zero, we use a numerical solve command (**fsolve** in Maple).

```
> fsolve(d=0,x);
```

$$0.8177316739$$

We can now begin to sketch the graph. Drawing the two asymptote lines ($y = 1$ and $x \approx 0.8$) and plotting the three x-intercept points brings us to the question: What do we need to know to complete the sketch? That is, how does the curve approach the asymptotes? (Is $f(x)$ positive and thus approaching positive infinity or is $f(x)$ negative and thus approaching negative infinity?) If we knew the intersection points (and their multiplicities) of the curve with the horizontal asymptote ($y = 1$), we could complete the sketch. Thus we solve $f(x) = 1$ for x.

```
> solve(f(x)=1,x);
```

$$0.3$$

We can now complete the graph of f. (The reader should explain how a global sketch of f can be drawn from the existing information.)

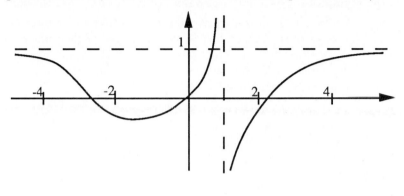

$\triangle$

Example 1 illustrates the fact that finding zeros is the major computational aspect involved in sketching the graph of a rational function. A CAS is a powerful tool for finding zeros. For example, one can do any of the following:

1. Obtain zeros by first factoring an expression using a CAS factor command

2. Use a CAS (rational) solve command to obtain the zeros in symbolic form

3. Use a CAS (numerical) solve command to obtain the zeros in decimal form

4. Use a root finding algorithm (e.g., the Bisection Algorithm)

5. Approximate the zeros from a plot

A combination of these methods is usually used in finding zeros. In general for polynomials of degree greater than 3, CASs can neither completely factor nor find all the zeros in symbolic form; thus methods 1 and 2 above have limited applicability. Numerical solve commands (e.g., **fsolve** in Maple) usually give just one zero. (However, Mathematica has a command to find all roots of a polynomial: NRoots$[p(x) == 0, x]$. MACSYMA has a similar command, **allroots**.) Thus it is necessary to first determine how many real zeros exist and, if there are more than one, to isolate the zeros by intervals before applying a numerical solve command. (Apply the numerical solve command to each interval.) The easiest way to determine the number of zeros and isolate them in intervals is to look at a plot of the expression. Since root finding algorithms usually require isolating a root in an interval, plotting is often a first step in method 4 above as well as in method 3.

Example 2

Sketch the graph of $f(x) = \dfrac{2x^5 + x^4 - 6x^3 - 3x^2 - 10x - 5}{x^5 - 2x^2 + 3x + 1}$.

We need to find the x-intercepts, the asymptotes, and where the graph intersects the asymptotes. Since for large values of x, $f(x) \approx \dfrac{2x^5}{x^5}$, we can tell by inspection that $y = 2$ is a horizontal asymptote. To find the x-intercepts, we need to find the zeros of the numerator. Since the degree of the numerator is 5, we know that there exists 1, 3, or 5 real zeros. To find the zeros, we will define f, extract the numerator, plot the numerator to determine the number of real zeros, and then use a solve command.

```
> f:=proc(x)(2*x∧5+x∧4-6*x∧3-3*x∧2-10*x-5)/(x∧5-2*x∧2+3*x+1)end;
> n:= numer(f(x));
```

$$n := 2x^5 + x^4 - 6x^3 - 3x^2 - 10x - 5$$

The plot of n over $[-5, 5]$ suggests that the numerator is negative for $x < -3$ and positive for $x > 3$. The result for $x < -3$ can be confirmed analytically by noting that the first three terms factor, i.e., $2x^5 + 4x^4 - 6x^3 = x^3(2x+3)(x-2)$, into a product of three factors each of which is negative for $x < -3$ and the sum of the last three terms is also negative for $x < -3$ as seen by writing them as $-x(3x - 10) - 5$. We leave the analytical verification that the numerator is positive for $x > 3$ to the student.

Thus we will plot over the interval $[-3, 3]$.

```
> plot(n,x=-3..3);
```

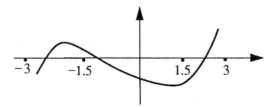

From this plot we see that there are three x-intercepts.

```
> fsolve(n=0,x);
```

$$-2.047579645, \quad -0.5000000000, \quad 2.047578645$$

To find the vertical asymptotes, we apply a similar type of analysis to the denominator.

```
> d:= denom(f(x));
```

$$d := x^5 - 2x^2 + 3x + 1$$

We now plot d over $[-2, 1]$. (Why choose this interval?)

```
> plot(d,x=-2..1);
```

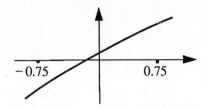

$$-0.75 \qquad 0.75$$

From this plot we see that there is only one zero and thus f has only one vertical asymptote.

```
> fsolve(d=0,x);
```

$$-0.2803562459$$

Now to find where the graph intersects the horizontal asymptote. That is, find the solutions of $f(x) = 2$ or, in terms of numerator and denominator, the solutions of $n = 2d$. Thus we will plot $n - 2d$ to determine the number of solutions and then solve for their values.

```
> plot(n-2*d);
```

Replotting to adjust the interval shows that there are 2 solutions.

```
> fsolve(n-2*d=0,x);
```

$$-0.4015238114, \qquad 6.275278152$$

We now have all the information needed for a global sketch of the desired graph.

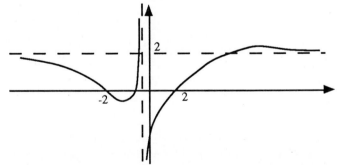

The reader should explain why this (global) sketch differs from a CAS plot. (Section 3.5 discusses how to use a CAS to find extremal values.) △

Exercises

1. Find all the real zeros of $f(x) = x^4 - 3x - 5$.

2. Find all the real zeros of $\sin(x^2) = x^2 + 2$.

In exercises 3–9, sketch your final graph by hand. Label all asymptotes and points of intersection.

3. Sketch the graph of $f(x) = \dfrac{x^2 - 1}{x^3 - 4x + 2}$.

4. Sketch the graph of $f(x) = \dfrac{x^3 + 2x^2 - x - 2}{x^4 - 3x - 5}$.

5. Sketch the graph of $f(x) = \dfrac{x^4 - 3x^2 - 5}{x^5 - 2x^2 + 3x + 1}$.

6. Sketch the graph of $f(x) = \dfrac{2x^4 + 3x}{x^4 + x - 4}$.

7. Sketch the graph of $f(x) = \dfrac{\sin(x^2) - x^2 - 2}{x^4 - 3x - 5}$.

8. Sketch the graph of $f(x) = \dfrac{(x - 1)^3}{x^3 - 7x + 3}$.

9. Sketch the graph of

$$f(x) = \frac{x^6 - 18x^4 + 4x^3 + 93x^2 - 36x - 108}{x^7 + 6x^6 - 7x^5 - 102x^4 - 117x^3 + 378x^2 + 891x + 486}.$$

10. Sketch the graph of $f(x) = \dfrac{x^5 - 8x^4 + 11x^3 + 56x^2 - 180x + 144}{x^3 - 6x^2 + 3x + 10}$.

3.5   Extremal Values

Prerequisites

- Second derivative test for extremal values

- Section on Sketching Rational Functions

Discussion

Determination of the extremal values (local and global) of a function is one of the most important applications of the calculus. Good graphing packages have simplified finding extremal values to the point where in many situations acceptable approximate answers can be obtained by plotting, zooming, and then digitizing the appropriate points. There are at least four reasons, however, for understanding the analytical approach.

(a) A graphical solution is an approximate solution. When the exact solution is required, an analytical approach is needed.

(b) Extremal value work for functions of three or more variables requires an analytical approach. (We can not plot in four or higher dimensions.) The analytical approach for functions of several variables is a natural generalization of the approach for functions of a single variable.

(c) A plot does not provide any information about the function outside of the domain of the plot. Thus one needs to (analytically) determine the domain for the plot.

(d) A plot may not be sufficiently clear. For example, consider a plot of $f(x) = \sin(\frac{1}{x})$ over an interval containing the origin.

Thus we will consider the analytical approach (with help from our graphics). Basically there are two steps involved. The first is to determine the critical points of the function, and the second is to analyze the behavior of the function at the critical points as well as at the endpoints of the domain. Critical points are the zeros of the derivative or the points at which the derivative is not defined. Thus the first step in determining extremal values consists of finding zeros of the function and determining the domain of the function, the function being the derivative. The major tools for carrying out the second step include graphing, first derivative test, second derivative test, and physical considerations of the situation being modeled.

In this section, we consider extremal value problems in which the computations required exceed the feasibility or practicality of pencil and paper calculations. Graphing will be used to "guide" the analysis.

Example

We will determine the extremal values of

$$f(x) = \frac{x(x-1)(x+3)}{(x+1)^2(x-2)(x-4)}$$

Step 1 is to find the critical values. Thus we need to find the derivative of f. We could do this by hand; however, a CAS will do it much faster (and probably more accurately). So, first we define f

```
> f:= proc(x) (x*(x-1)*(x+3))/((x+1)^2*(x-2)*(x-4)) end;
```

Now we define a function df as the derivative of f

```
> df:= proc(y) subs(x=y,diff(f(x),x)) end;
```

(Note that we cannot just write df := diff(f(x),x); since diff returns an expression rather than a function. Maple and Mathematica *do* have differentiation *operators* which return a function. We do not use the differentiation operator since it is not commonly found in CASs, but you might try using it.)

To find the zeros of df, solve the equation $df(x) = 0$. That is

```
> solve(df(x) = 0,x);
```

The lack of response indicates that it is not capable of solving for the zeros of df. Thus an alternative approach is required.

Let us sketch the graph of f on an interval containing the points of interest, such as -3 (a zero), 4 (an asymptote), etc. (See the Section on Sketching Rational Functions if you want to do it by hand rather than using the computer's plotting abilities.) Since we know that $f(x)$ becomes unbounded as x approaches $x = -1, 2,$ or 4 (vertical asymptotes), we expect the vertical scale to distort the graph. Thus we restrict the y range to $[-5, 5]$ in order to have a meaningful picture of the graph near the x-axis.

> plot(f(x),x=-5..5, y=-5..5);

This returns (with asymptote lines added by hand)

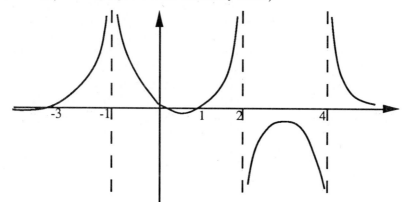

The x-axis is a horizontal asymptote and there are vertical asymptotes at $x = -1$, 2, and 4. Furthermore the vertical asymptotes at $x = 2$ and 4 result from factors of multiplicity one while the asymptote at $x = -1$ results from a factor of multiplicity two. There are x-intercepts at $x = -3$, 0, and 1. Since all of the x-intercepts are given by factors of multiplicity one, the curve crosses the x-axis at each of the intercepts.

It is clear that f has a local maximum between 2 and 4, a local minimum between 0 and 1, and possibly another local minimum for some value of $x < -3$, and no global extrema. Note that the sketch has accomplished Step 2 in the procedure for determining extremal values. The sketch also indicates that the important critical values are in the intervals $[2.25, 3.75]$, $[.25, .75]$, $[-10, -3]$ (Why?). We now use the Bisection Algorithm to approximate these critical values.

If your CAS has a built-in numerical zero solving function, such as the Bisection Algorithm or fsolve in Maple, it can be used. If none is available, the hand-crafted algorithm "bisect" in the section on the Bisection Algorithm can be used.

> bisect(df,2.25,3.75,0.001);

The output rounded to three decimal places is 2.730. Evaluating $f(2.730)$

```
> f(2.730);
```

yields -2.098. Thus f has a local maximum at $(2.730, -2.098)$.

Similar analyses of the intervals $[.25, .75]$ and $[-10, -3]$ show local minimums at $(0.469, -0.074)$ and $(-6.411, -0.063)$.

We verify these results analytically by applying the second derivative test. Denoting the second derivative by ddf, we have

```
> ddf:= proc(y) subs(x=y,diff(df(x),x)) end;
```

Evaluating the second derivative at the critical values yields

$$ddf(2.730) \quad = \quad -4.493 \ \Longrightarrow \ \text{concave downward} \ \Longrightarrow \ \text{local maximum}$$
$$ddf(0.469) \quad = \quad +0.498 \ \Longrightarrow \ \text{concave upwards} \quad \Longrightarrow \ \text{local minimum}$$
$$ddf(-6.411) \ = \quad +0.002 \ \Longrightarrow \ \text{concave upwards} \quad \Longrightarrow \ \text{local minimum}$$

Thus we have found the extrema.   △

Exercises

Determine the extremal values, local and global. Verify your results by applying the second derivative test.

1. $f(x) = \dfrac{(x-4)}{x^2(x-2)}$

2. $f(x) = \dfrac{(x-5)(x+1)(x-2)}{(x+2)^2(x-3)^2}$

3. $f(x) = \dfrac{x^3}{(x+2)(x-1)(x-4)}$

4. $f(x) = \dfrac{\sin(x)}{(x+1)(x-1)^3}$

5. (Exploratory) The objective of this exercise is to describe the path traced by the critical point of the graph of $y = ax^2 + bx + c$ when only one coefficient is varied.

    (a) Consider (*) $y = ax^2 + 2x + 1$. Superimpose the graphs of five quadratics obtained from (*) by assigning five different values to a. Locate (visually) the critical points of the five graphs.

    (b) Repeat step (a) as many times as necessary in order to conjecture a description of the path traced by the critical points when the value of the a coefficient is varied.

    (c) "Check out" your conjecture by repeating steps (a) and (b) with another quadratic in which the b and c coefficients are fixed.

(d) Give an algebraic verification of your conjecture. Hint: Consider the general quadratic, $y = ax^2 + bx + c$, with coefficients b and c fixed. Determine the relation between a and b at the critical point. Solve for b in terms of a and substitute this value of b into the general quadratic. The result will be the algebraic expression for the curve of the critical point when a is varied. (Why?)

(e) Repeat steps (a)–(d) for a quadratic in which only the b coefficient is varied.

(f) Repeat steps (a)–(d) for a quadratic in which only the c coefficient is varied.

6. (Exploratory) Describe the curves of the critical points of the graph of a cubic, $y = ax^3 + bx^2 + cx + d$, when only one coefficient is varied. Follow the procedure outlined in Exercise 5 for quadratics.

7. (Exploratory) Describe the curves of the critical points of the graph of $y = \arcsin(bx + c)$, $-\pi \le x \le \pi$, when only one of the parameters is varied. Follow the procedure outline in Exercise 5.

8. (Exploratory) For a differentiable function f, the critical points are the zeros of f'. Geometrically, the zeros of f' are the points where the graph of f' touches the x-axis. The following figures show the four ways that a graph can touch the x-axis. For each of the four ways (a-d):

(a) Conjecture an implication for f (with respect to extreme values). For example, does the critical point imply the existence of an extreme value? If so, is the extreme value a relative minimum or a relative maximum? Give a written verification of your conjecture.

(b) Conjecture whether the second derivative of f is negative, positive, or zero at the critical point. Give a written verification of your conjecture.

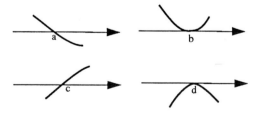

3.6   Fixed Points

Prerequisites

- The Intermediate Value Theorem

- The Mean Value Theorem

- The section on the Bisection Algorithm

- The section on Convergence of Sequences

Discussion

In this section we will discuss solutions of the equation

$$f(x) = x$$

Such an x is called a *fixed point* of f since f leaves x fixed. Many problems can be reduced to a fixed point problem. For example, suppose you wish to solve

$$f(x) = b$$

If we let $F(x) = f(x) + x - b$, then

$$F(x_0) = x_0 \iff f(x_0) + x_0 - b = x_0 \iff f(x_0) = b$$

Thus the solution to $f(x) = b$ is the fixed point of F.

We will first consider some sufficient conditions which guarantee that f has a fixed point. Many simple functions have no fixed point. For example, consider

$$f(x) = x + 1$$

If $f(x_0) = x_0$, then $x_0 + 1 = x_0$ or $1 = 0$, which is impossible. There is one easy theorem which guarantees the existence of a fixed point.

Theorem 3.6.1 *The Fixed Point Existence Theorem. Let $f : [a, b] \to [a, b]$ be continuous. Then f has a fixed point in $[a, b]$.*

Proof: Consider the graph of f. If $f(a) = a$ or $f(b) = b$, then f has a fixed point. If neither of these cases holds, then $f(a) > a$ and $f(b) < b$ (Why?) and the graph of f looks like:

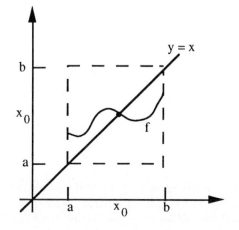

The graph of f must cross the line $y = x$ at some point x_0, at which point $f(x_0) = y = x_0$. (To see this rigorously, we can apply the Intermediate Value Theorem (IVT) to $g(x) = f(x) - x$. The function g is continuous, $g(a) > 0$, and $g(b) < 0$. Thus by the IVT there is an x_0 between a and b such that $g(x_0) = f(x_0) - x_0 = 0$.)   □

This theorem only gives sufficient conditions for the existence of a fixed point. A function which doesn't satisfy the hypothesis may still have a fixed point.

Example 1

Consider the function

$$f(x) \;=\; \cos(x) + \frac{1}{4}$$

This function does not satisfy the Fixed Point Existence Theorem's hypothesis on the interval $[0,1]$ since $f(0) = 1.25$ is not in $[0,1]$, but since

$$g(x) \;=\; f(x) \;-\; x \;=\; \cos(x) \;+\; \frac{1}{4} \;-\; x$$

is continuous and satisfies $g(0) > 0$ and $g(1) < 0$, we know that at some point x_0 between 0 and 1, $g(x_0) = 0$, and thus $f(x_0) = x_0$. We can use the Bisection Algorithm to find the fixed point.   △

There may be many fixed points. For example, for the identity function $f(x) = x$, every point is a fixed point. The following result gives sufficient conditions for the existence of a *unique* fixed point.

Theorem 3.6.2 *The Fixed Point Existence and Uniqueness Theorem.*   Let $f : [a,b] \rightarrow [a,b]$ *be continuous. If there is a bound $M < 1$ such that $|f'(x)| \le M$ for all x in (a,b), then f has a unique fixed point in $[a,b]$.*

Proof: By the previous theorem, we know that f has a fixed point. Suppose, if possible, that f has two fixed points, $p \ne q$, in $[a,b]$. Then by the Mean Value Theorem there is an r between p and q such that

$$f(p) \;-\; f(q) \;=\; f'(r)(p \;-\; q)$$

thus

$$|p \;-\; q| \;=\; |f(p) \;-\; f(q)| \;=\; |f'(r)||p \;-\; q| \le M|p \;-\; q| \;<\; |p \;-\; q|$$

which is impossible.   □

The above theorem provides a way to find an approximation to the fixed point other than by the Bisection Algorithm. The new method, called *Fixed Point Iteration*, sometimes converges more rapidly than the Bisection Algorithm.

Theorem 3.6.3 *Fixed Point Iteration Theorem and Algorithm. Suppose that* $f : [a, b] \rightarrow [a, b]$ *is continuous and that there is a bound* $M < 1$ *such that* $|f'(x)| \leq M$ *for all* x *in* (a, b). *Then for any* x_0 *in* $[a, b]$ *the sequence of points defined by*

$$x_{n+1} = f(x_n), \quad n \geq 0$$

converges to the unique fixed point p, *and the error satisfies*

$$|p - x_n| \leq M^n |b - a|$$

Proof: As in the previous proof, by the Mean Value Theorem,

$$|p - x_n| = |f(p) - f(x_{n-1})| \leq M |p - x_{n-1}|$$

Applying the same argument again, we have

$$|p - x_{n-1}| \leq M |p - x_{n-2}|$$

or

$$|p - x_n| \leq M^2 |p - x_{n-2}|$$

Continuing for n steps we have

$$|p - x_n| \leq M^n |p - x_0| \leq M^n |b - a|$$

Since $M < 1$, the right hand side converges to 0 as $n \rightarrow \infty$ and thus $x_n \rightarrow p$. $\square$

Note that if $M < \frac{1}{2}$, then convergence will be faster than the Bisection Algorithm, which reduces the size of the interval by a factor of $\frac{1}{2}$ at each step.

Example 2

Suppose we wish to find a solution to

$$\cos(x) = 3x$$

with an accuracy of 0.01. If we solve for x and let $f(x) = \frac{\cos(x)}{3}$, then the problem is equivalent to finding a fixed point of f. Since $f : [0, \frac{\pi}{2}] \rightarrow [0, \frac{\pi}{2}]$ is continuous and

$$|f'(x)| = |\frac{-\sin(x)}{3}| \leq \frac{1}{3} \text{ on } (0, \frac{\pi}{2})$$

the Fixed Point Iteration method will converge, with an error of at most $\frac{1}{3^n}(\frac{\pi}{2} - 0)$ after n iterations. For the error to be less than 0.01, we need

$$\frac{\pi}{2 \cdot 3^n} \leq 0.01$$

or

$$3^n \geq 50\pi \approx 157$$

so $n = 5$ iterations will be sufficient.

We will let $x_0 = 0$, so that

$$x_1 = .3333333333$$
$$x_2 = .3149856487$$
$$x_3 = .3169336083$$
$$x_4 = .3167318460$$
$$x_5 = .3167527996$$

Thus the solution is $x = 0.31675$ with an error of at most 0.01.   △

The iteration procedure can be put into a loop, using the **for** command (see the subsection on Control Structures in the Getting Started section). Thus, to iterate 5 times:

```
> f:= proc(x) evalf(cos(x)/3) end;
> x:= 0;
> for n from 1 to 5 do
≫         x:= f(x);
≫ od;
```

Exercises

1. Give an example of a function $f : \Re \to \Re$ whose only fixed point is 1.

2. Consider the *quadratic map* defined by $f(x) = \lambda x(1 - x)$. As λ increases from 0, how do the number and location of the fixed points change? Hint: Graph f for increasing values of λ and use the idea in the proof of the Fixed Point Existence Theorem (i.e., note where the graph of f crosses the line $y = x$).

3. We wish to find the smallest positive solution to

$$x^3 - 5x + 1 = 0$$

   (a) What is the function f for the corresponding fixed point problem, $f(x) = x$?

   (b) What is the interval $[a, b]$ with $f([a, b]) \subseteq [a, b]$ and $|f'(x)| \leq M < 1$ on $[a, b]$?

   (c) What is the maximum number of iterations needed for an error of at most 0.001?

   (d) What is the solution?

4. Let $f : [0, 1] \to [0, 1]$ be defined by $f(x) = 1 - x$.

   (a) Find the fixed point(s) of f, if it has any.

   (b) Determine the sequence of iterations of $f(x)$, i.e., $x_{n+1} = f(x - n)$, $n \geq 0$. Explain the behavior of this sequence and why it does or does not converge to the fixed point.

5. Show that $f(x) = \dfrac{\sin(x)}{2} + \pi$ has a unique fixed point on $[0, 2\pi]$. Then

(a) Determine the (theoretical) maximum number of iterations required to approximate the fixed point with an error of at most 0.01.

(b) Form a table of iterations (i.e., use a **for** loop) for the number of iterations determined in part (a).

(c) Using your table in part (b), determine the least number of iterations required to obtain an approximation to the fixed point that is accurate to within 0.01.

(d) Compare the "heuristic" number of iterations needed [from part (c)] to the theoretical number [from part (a)]. Explain why the heuristic number can be considerably less than the theoretical number.

3.7   Iteration and Chaos

Prerequisites

- The section on Recursive Functions

- The section on Convergence of Sequences

- The section on Fixed Points

Discussion

When a function is composed with itself, we say that the function is *iterated*. A computer or calculator with a function key affords an interesting way to generate recursive sequences through iteration. The following example illustrates the process with the square root function.

Example 1. Iterating the Square Root Function

Let x_0 be a positive number. Let $f(x) = \sqrt{x}$. We will generate the sequence $f(x_0)$, $f(f(x_0))$, $f(f(f(x_0)))$, etc. The sequence can be defined recursively by letting x_0 be the initial value (the base case) and letting $x_n = f(x_{n-1})$ (the inductive step). With a calculator enter x_0 and then press the square root key repeatedly. The screen displays form a recursive sequence. With our CAS doing the computations on an initial value of $x_0 = 2$,

```
> x:= 2;
> for n from 0 to 11 do
>>        print(x);
>>        x := sqrt(x);
>> od:   # colon to avoid unwanted output.
```

we obtain

  2, 1.414213562, 1.189207115, 1.090507733, 1.044273782, 1.021897149,
  1.010889286, 1.005429901, 1.002711275, 1.00135472, 1.000677131,
  1.000338508, ...

In this example, 2 is the base case and applying the square root function is the inductive step.   △

The elements in the recursive sequence of the previous example seem to be converging (i.e., getting close) to one. Is this result dependent or independent on the choice of 2 for the base case? The reader should experiment, redoing the previous example using other values for the base case and then form a conjecture concerning the long term behavior of these sequences.

The iteration process for generating a recursive sequence has an interesting graphical interpretation. We will illustrate it for the previous example. Draw the graphs of the square root function, $y = f(x) = \sqrt{x}$, and the identity function, $y = i(x) = x$ on the same set of axes. Starting with the base case, x_0, move vertically to $(x_0, f(x_0))$ on the square root function curve and then move horizontally to $(f(x_0), f(x_0))$ on the identity function curve. This point represents the first iteration. Now repeat this two step process, i.e., move vertically (from $(f(x_0), f(x_0))$) to $(f(x_0), f(f(x_0)))$ on the square root function curve and then move horizontally to $(f(f(x_0)), f(f(x_0)))$ on the identity function curve. This point represents the second iteration. The continuation of this process yields what is called a "cobweb" graph.

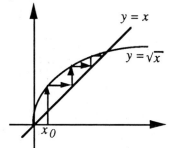

The reader should give a geometric argument based on the above cobweb graph that the sequence in the previous example does converge to one.

Example 2. Iterates of $f(x) = -x^3$

Consider the function $f : \Re \to \Re$ defined by $f(x) = -x^3$. Let u_n, v_n, and w_n be the three sequences obtained by iterating f on the base cases $1.1, -1$, and 0.8. Thus $u_0 = 1.1$ and $u_n = f(u_{n-1})$ for $n \geq 1$; $v_0 = -1$ and $v_n = f(v_{n-1})$ for $n \geq 1$; and $w_0 = 0.8$ and $w_n = f(w_{n-1})$ for $n \geq 1$, respectively. Computing the first few elements of these sequences yields:

  $\{u_n\} = \{1.1, -1.331, 2.357948, -13.109994, 2253.240236, \ldots\}$

$$\{v_n\} = \{-1, 1, -1, 1, -1, 1, -1, ...\}$$
$$\{w_n\} = \{0.8, -0.512, 0.13421773, -0.00241785, 0.00000001, ...\}$$

The "cobweb" graphs of the three sequences are shown in the following figure.

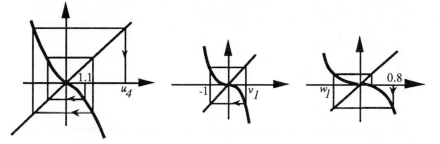

$\{u_n\}$ is an example of an *unbounded, divergent* sequence. Note that the elements of $\{u_n\}$ become very large in both the positive and negative sense and thus the elements will not eventually cluster about any number.

$\{v_n\}$ is an example of a *bounded, divergent* sequence. $\{v_n\}$ is also an example of a *periodic* sequence. Note that the elements of a periodic sequence will never eventually cluster about a *single* number.

$\{w_n\}$ is an example of a *convergent* sequence. Even though the elements alternate in sign, they seem to be closely *approximating* the number zero. Thus, given an approximation w_n, a better approximation is obtained by applying f to w_n to obtain $w_{n+1} = f(w_n)$. $\triangle$

Exercises

1. Let $\{s_n\}$ be the recursive sequence obtained by iterating the square root function applied to 2.

    (a) Determine an expression for s_n.

    (b) Show that $\{s_n\}$ is a decreasing sequence by showing that $\frac{s_{n+1}}{s_n} < 1$.

    (c) What theorem on convergence guarantees that $\{s_n\}$ is convergent?

    (d) Use a computer or calculator to find the smallest value of n such that s_n approximates its limit with an error of at most 0.000005.

2. Find several iterates (x_1 through x_4) of $f(x) = 2x$ with $x_0 = 1$, expressed as decimal integers. If $f^n(x) = f \circ f \cdots \circ f(x)$ is the nth iterate, can you find a formula for $f^n(x)$?

3. Repeat Exercise 2 with $f(x) = x + 2$ and $x_0 = 1$.

4. Plot the graphs of $f(x) = x^2 - 2$ and $y = x$ on the same set of axes. Using a cobweb graph, describe the behavior of the iterates of $f(x_0)$ for

(a) $x_0 = -2.5$

(b) $x_0 = 0$

(c) $x_0 = 2.5$

5. For each of the following three functions: choose various initial values x_0, compute iterates as described in Exercise 1, and then graphically display the results with cobweb graphs. Do you find any consistent behavior?

   (a) $f(x) = \sin(x)$

   (b) $f(x) = \cos(x)$

   (c) $f(x) = 2^x$

6. Generate three sequences by iterating the function $g : \Re \to \Re$ defined by $g(x) = x^3$ on the base cases: 1.3, −1, 0.9. Using a computer or calculator, compute the first few values of each of the sequences. Sketch the cobweb graphs to illustrate the behavior of the sequences and then intuitively analyze the results (as was done in the above example on $f(x) = -x^3$).

7. Consider the recursive sequence $\{s_n\}$ obtained by iterating the function $f : \Re \to \Re$ defined by $f(x) = x^2$ applied to the number x_0.

   (a) Draw two cobweb graphs for $\{s_n\}$, one with $x_0 = -0.9$ and one with $x_0 = 1.1$.

   (b) Iterate at least 8 times for each starting value, $x_0 = -2, -0.6, 0.9, 1.1$.

   (c) Using the results of parts a and b, make a conjecture concerning the convergence or divergence of $\{s_n\}$ with respect to the size of the absolute value of x_0.

8. Let $\{s_n\}$ be the sequence obtained by iterating the function $f(x) = x^2$ applied to x_0, $-1 < x_0 < 1$.

   (a) Determine an expression for s_n.

   (c) Show that $\{s_n\}$ is a decreasing sequence by showing that $\frac{s_{n+1}}{s_n} < 1$.

   (c) What theorem on convergence shows that $\{s_n\}$ is convergent?

   (d) Use a computer or calculator to find the smallest value of n such that s_n approximates its limit with an error of at most 0.000005.

Example 3. Chaotic Behavior

This example is a simple model which can exhibit chaotic behavior. Let P_n be the number of insects alive in year n; P_0 is the initial population. The number of insects alive in year $n + 1$ is dependent upon the number alive in year n: insects alive in one year reproduce larva which survive the winter, but the adults die during the winter. A simple model is

$$P_{n+1} = aP_n - bP_n^2$$

Here $a > 0$ represents the birthrate: the higher the birthrate, the more insects there will be in the next generation. The $b > 0$ coefficient represents the effects of crowding, which is proportional to the square of the population size. By an appropriate choice of units we can set $a = b = \lambda$ and our model becomes

$$P_{n+1} = \lambda P_n(1 - P_n)$$

The function $f(x) = \lambda x(1 - x)$ is called the *quadratic map* and will appear in several exercises. In these units $0 < P_n < 1$. The behavior of P_n as λ increases from 0 to 4 is quite surprising. When $\lambda < 1$, the population dies out. We use our CAS to do the computation. Setting $\lambda = 0.5$ and $P_0 = 0.8$ we have

```
> f:= proc(x,l) l*x*(1-x) end;
> x:= 0.8;
> for n from 0 to 4 do
>>        print(n,x);
>>        x:= f(x,l);
>> od:   # colon to avoid unwanted output.
```

This produces

n	P_n
0	0.8
1	0.08
2	0.0368
3	0.0177
4	0.0087

For $3 < \lambda < 3.4$, P_n oscillates between two values. For $3.4 < \lambda < 3.57$, P_n oscillates between 4, then 8, then 16, ... values. When $\lambda > 3.57$, the behavior of P_n is chaotic: the long term behavior of P_n is extremely sensitive to the initial value P_0. For $0 < \lambda < 3$, the long term behavior is independent of the initial value. The reader will be asked to verify these statements in the exercises. $\triangle$

Exercises

9. This exercise examines the model of chaotic growth introduced in the above example.

   (a) Choose several values of λ between 0 and 1. Verify that P_n dies out as n increases.

   (b) Choose $\lambda = 2$, and then several other values of λ between 1 and 3. How does P_n behave as n increases? How is the behavior dependent upon the initial population P_0 or the value of λ?

(c) Choose several values of λ between 3 and 3.57. How does P_n behave as n increases? The behavior should be periodic. How is the behavior dependent upon the initial population P_0 or the value of λ?

(d) Choose several values of λ greater than 3.57. How does P_n behave as n increases? How is the behavior dependent upon the initial population P_0 or the value of λ?

10. Some of the behavior of the quadratic map, $f(x) = \lambda x(1 - x)$, $x \in [0, 1]$, can be explained by examining its fixed points (see the Section on Fixed Points).

(a) Graph the quadratic map for several values of λ. Note that it is a parabola with the axis along $x = 0.5$. Any fixed point must lie on the intersection of $y = x$ with $y = f(x)$. What fixed point does f always have?

(b) Find the other fixed point of f in [0,1] as a function of λ. How is this fixed point related to the behavior investigated in Exercise 9?

11. Consider the quadratic map, $f(x) = \lambda x(1 - x)$, $x \in [0, 1]$.

(a) Draw the cobweb graph for f when $\lambda = 0.5$. Does it display the behavior discussed in the example?

(b) Draw the cobweb graph for f when $\lambda = 2$. Does it display the behavior discussed in Exercises 9(b) and 10?

Example 4. Classification of Fixed Points

Fixed points can be classified as *attracting* or as *repelling*, depending on the behavior of iterates near the point. A fixed point is attracting if iterates of nearby points tend towards the fixed point; the fixed point is repelling if iterates of nearby points tend away from the fixed point.

Consider $f(x) = x^2$. This function has two fixed points, 0 and 1. Let x_0 be a point within the interval $(-1, 1)$ and not equal to 0. Then the sequence of iterates

$$x_n = x_0^{2^n}$$

converges to the fixed point 0. Let x_0 be a point within the interval $(0, 2)$ and not equal to 1. Then the sequence of iterates moves away from the fixed point 1. This behavior can also be seen by considering the cobweb graphs for the two fixed points:

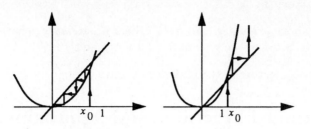

or by numerically iterating some values:

```
> x:= 0.5;
> for n from 0 to 10 do
≫        print(x);
≫        x:= x∧2;
≫ od:
```

The values tend away from 1, a repelling fixed point, and towards 0, an attracting fixed point.

A fixed point may be neither attracting nor repelling. Consider $f(x) = x$. Every point is a fixed point, and so no fixed point either attracts or repels nearby points. $\triangle$

Exercises

12. Examine the following functions, classifying their fixed points.

   (a) $f(x) = \dfrac{x}{3}$

   (b) $f(x) = \dfrac{x}{2}$

   (c) $f(x) = \dfrac{3x}{2}$

   (d) $f(x) = 2x$

13. Examine the following functions, classifying their fixed points.

   (a) $f(x) = x^3$

   (b) $f(x) = x^4$

   (c) $f(x) = -x^3$

   (d) $f(x) = -x^4$

14. Examine the following functions, classifying their fixed points.

   (a) $f(x) = \sin(x)$

   (b) $f(x) = \cos(x)$

(c) $f(x) = \exp(x)$

15. From the above examples, and from others if necessary, what relation do you observe between the type of fixed point p and $f'(p)$?

16. Prove your conjecture in the previous exercise.

3.8   Continued Fractions and Approximations

Prerequisites

- The section on Elementary Graphing

- The section on Fixed Points

- The section on Convergence of Sequences

Discussion

This section, based on a paper by Tony Crilly [3], explores continued fractions. Continued fractions can be derived from the Greek ideas on measurement and provide a way to represent or approximate numbers. The usual base 10 representation is arbitrary, a result of having 10 fingers. The continued fraction representation is more intrinsic, and provides a "best" approximation to an irrational number.

The continued fraction representation of a positive number a has the form

$$a = a_1 + \cfrac{1}{a_2 + \cfrac{1}{a_3 + \cfrac{1}{a_4 + \cfrac{1}{\ddots}}}}$$

where $a_1, a_2, a_3, a_4, \ldots$ are a sequence of positive integers called the *spectrum* of a. (If $a < 1$, then $a_1 = 0$.) The continued fraction representation of a is written as $[a_1, a_2, a_3, a_4, \ldots]$ in "condensed" format. Clearly a is rational if the continued fraction representation terminates, i.e., the spectrum is finite. The converse is also true. The rational number

$$r_n = a_1 + \cfrac{1}{a_2 + \cfrac{1}{a_3 + \cfrac{1}{\ddots \atop a_{n-1} + \cfrac{1}{a_n}}}} = [a_1, a_2, a_3, \ldots, a_{n-1}, a_n]$$

is called the *nth-order convergent* or *nth-order approximant* of a. If a is irrational, then $\lim_{n \to \infty} r_n = a$.

Example 1

As an example, consider $a = 5/3$. First we truncate $5/3$ to obtain the integer part 1, so that:

$$\frac{5}{3} = 1 + \frac{2}{3}$$

Then we take the reciprocal of the fraction to obtain the desired 1 in the numerator:

$$\frac{5}{3} = 1 + \frac{2}{3} = 1 + \frac{1}{\frac{3}{2}}$$

We repeat these two steps with the new fraction $3/2$. We stop when the new fraction is an integer.

$$\frac{5}{3} = 1 + \frac{2}{3} = 1 + \frac{1}{\frac{3}{2}} = 1 + \frac{1}{1 + \frac{1}{2}}$$

This is written more briefly as $[1, 1, 2]$.

Note that the continued fraction representation of $2/3$ is $[0,1,2]$ since the integer portion is zero. △

Computation of the Continued Fraction Representation

Let's apply these steps to an arbitrary positive number a. To obtain its integer portion, we use the *floor* or *greatest integer* function, $[a]$, which truncates a positive number to its integer part. (For example, $[2.9] = 2$.) The fractional portion is $a - [a]$. Thus we have

$$a = [a] + (a - [a])$$

As above, we then take the reciprocal of the fractional part,

$$a = [a] + \frac{1}{\frac{1}{a - [a]}}$$

If we define f by

$$f(x) = \frac{1}{x - [x]}$$

for positive, non-integer values of x, then

$$a = [a] + \frac{1}{f(a)}$$

Repeating this process on the new value $f(a)$, we have

$$a = [a] + \cfrac{1}{[f(a)] + (f(a) - [f(a)])} = [a] + \cfrac{1}{[f(a)] + \cfrac{1}{f(f(a))}} = [a] + \cfrac{1}{[f(a)] + \cfrac{1}{f^2(a)}}$$

where

$$f^2(a) = f(f(a)) = \frac{1}{f(a) - [f(a)]}$$

Continuing we have

$$a = [a] + \cfrac{1}{[f(a)] + \cfrac{1}{f^2(a)}} = [a] + \cfrac{1}{[f(a)] + \cfrac{1}{[f^2(a)] + (f^2(a) - [f^2(a)])}}$$

$$= [a] + \cfrac{1}{[f(a)] + \cfrac{1}{[f^2(a)] + \cfrac{1}{f^3(a)}}}$$

where $f^k(a)$ is f composed with itself k times, not $f(a)$ to the kth power.

Thus the integers $[a]$, $[f(a)]$, $[f^2(a)]$, $[f^3(a)]$, ... are the spectrum of a and the continued fraction representation for a is $[[a], [f(a)], [f^2(a)], [f^3(a)], \ldots]$. As above, we stop when the new value, $f(a)$, is an integer. (Continuing would result in division by $f(a) - [f(a)] = 0$.)

We can use our CAS and a for loop to compute the continued fraction representation.

First we define our function, calling our computerized version cfv (for Continued Fraction Value or Computerized F Version):

```
> cfv:= proc(x) 1/(x-trunc(x)) end;
```

We can now check our work by determining the partial fraction representation of 5/3.

```
> a:= 5/3;
> for k from 1 to 10 do
≫          print( k, 'th elt of spec = ', trunc(a));
≫          # Leave loop if a is integer.
≫          if a = trunc(a) then break fi;
≫          a:= cfv(a);
≫ od;
```

This returns the values 1, 1, and 2 as the spectrum of 5/3.

Example 2

We will find the partial fraction representation of an irrational number, such as $\sqrt{2}$. First we convert $\sqrt{2}$ to a decimal (since *trunc* requires a numerical value):

```
> a:= evalf(sqrt(2));
```

Now we apply the above **for** loop. It returns the first 10 integers in the spectrum 1,2,2,2,2,2,2,2,2,2. Does this regularity persist? To check further, we should increase the accuracy of the decimal approximation,

```
> Digits:= 20;
```

and try again. The regularity persists. Indeed, the spectrum of $\sqrt{2}$ is $[1,2,2,2,...]$. That is,

$$\sqrt{2} = 1 + \cfrac{1}{2 + \cfrac{1}{2 + \cfrac{1}{2 + \cfrac{1}{\ddots}}}}$$

We can use our function f to see the regularity of the spectrum of $\sqrt{2}$:

$$\sqrt{2}, \qquad\qquad\qquad [\sqrt{2}] = 1$$
$$f(\sqrt{2}) = \tfrac{1}{\sqrt{2}-1} = \sqrt{2}+1 \qquad [f(\sqrt{2})] = 2$$
$$f(\sqrt{2}+1) = \tfrac{1}{\sqrt{2}-1} = \sqrt{2}+1 \qquad [f^2(\sqrt{2})] = 2$$

Thus if $k > 0$, $[f^k(\sqrt{2})] = 2$. We will prove a more general result after some investigation of our function f. $\triangle$

Some Properties of f

If we evaluate f at an integer value, then an error will occur, since $x - trunc(x)$ is zero. We want to modify the definition of f so that it is defined at integer values. What value should $f(x)$ have when x is an integer? We will graph f to see if the graph gives us any ideas:

```
> # We restrict the y values since they go to infinity.
> plot(cfv,x=0..4,y=0..4);
```

This returns:

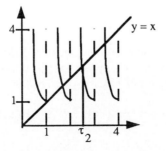

We can see that $f(x)$ approaches 1 as x approaches an integer from the left. Thus we modify the definition of f (and of the computerized version):

```
> cfv:= proc(x) if x = trunc(x) then 1
≫         else 1/(x-trunc(x)) fi end;
```

Now f is defined for all positive real numbers. It has an essential discontinuity at every positive integer and has period one, as is seen from the graph.

In actual computation on computers, which have only limited accuracy, there may be a problem with round-off errors when x is close to $trunc(x)$. So we will make a final modification to cfv to return 1 if x is close to $trunc(x)$. How close? If Digits is the number of digits of accuracy, then 10^(-Digits) is the smallest possible value. To be conservative, we will use half as many digits of accuracy. Thus we define cfv by

```
> cfv:= proc(x)
≫         if abs(evalf(x-trunc(x)))<10^(-Digits/2) then 1
≫         else 1/(x-trunc(x)) fi
≫ end;
```

The fixed points of f are the points of intersection of $y = f(x)$ and the line $y = x$. From the above graph, we see that there are infinitely many fixed points. To find them, we solve the equation

$$x = f(x) = 1/(x - trunc(x)) = 1/(x - [x])$$

or

$$x^2 - [x]x - 1 = 0$$

If we restrict x to an interval $(n, n + 1)$ $(n \geq 1$ an integer), then $[x] = n$ is constant and our equation becomes

$$x^2 - nx - 1 = 0$$

which, by the quadratic formula (or by our CAS), has one solution in $(n, n + 1)$,

$$\tau_n = \frac{1}{2}(n + \sqrt{n^2 + 4})$$

Note that $\tau_1 = \frac{1}{2}(1 + \sqrt{5})$ is the "golden mean". What happens to the sequence $\{\tau_n\}$ as n approaches infinity? From the graph, it appears that τ_n is close to n when n is large. We can use our CAS to verify this:

```
> limit(1/2*(n+sqrt(n^2 +4))-n,n=infinity);
```

returns 0.

A Theorem of Euler's

Using f we can obtain the following theorem due to Euler:

Theorem 3.8.1 *Let n be a positive integer. Then the continued fraction representation of $\sqrt{1+n^2}$ is*

$$[n, 2n, 2n, 2n, \ldots]$$

Proof: The continued fraction representation of $a = \sqrt{1+n^2}$ is

$$[[a], [f(a)], [f^2(a)], \ldots]$$

Now, $[a] = n$ and

$$f(a) = \frac{1}{\sqrt{1+n^2} - n} = \sqrt{1+n^2} + n$$

But

$$f(a) = \frac{2}{2}(\sqrt{1+n^2} + n) = \frac{1}{2}(\sqrt{4 + (2n)^2} + 2n) = \tau_{2n}$$

which is a fixed point of f. Thus for any positive integer $k > 1$,

$$[f^k(a)] = [f(a)] = [\sqrt{n^2 + 1} + n] = 2n$$

This gives the continued fraction representation of a as $[[n], [2n], [2n], \ldots]$. $\square$

As a special case, we have the continued fraction representation

$$\sqrt{2} = \sqrt{1+1} = [1, 2, 2, 2, 2, \ldots]$$

Best Approximations

When is a fraction p/q ($p > 0$ and $q > 0$) a *best* approximation to a positive number a? Clearly, if a is irrational, there is a rational number closer to a than p/q. However, such a closer approximation might require a larger denominator. Thus a natural definition is: p/q is a best approximation to a relative to n if p/q is closer to a than any other fraction p'/q' with $0 < q' \le n$. Thus, to get a better approximation than p/q, the denominator must be increased. Continued fractions give best approximations: if a is an irrational number with nth-order approximant $p/q = r_n = [a_1, a_2, \ldots, a_n]$, then p/q is a best approximation to a relative to q.

Example 3

Consider $a = \sqrt{2}$. We know that the continued fraction representation of $\sqrt{2}$ is $[1, 2, 2, 2, \ldots]$. We can find the 3rd-order approximant by hand:

$$1 + \cfrac{1}{2 + \cfrac{1}{2 + \cfrac{1}{2}}} = \frac{17}{12}$$

Thus $17/12$ is a best approximation to $\sqrt{2}$ relative to 12. The difference $17/12 - \sqrt{2} \approx 0.0025$. If we use the base 10 approximations to $\sqrt{2}$, 1, $1.4 = 14/10$, $1.41 = 141/100$, with errors 0.4, 0.01, and 0.004 respectively, we see that in the decimal expansion the denominator must increase above 100 to obtain an approximation better than $17/12$. $\triangle$

How accurate is the partial fraction approximation? If the nth-order approximant to an irrational positive number a is p_n/q_n, then $a - p_n/q_n < 1/q_n^2$. Thus in our previous example, $p_n/q_n = 17/12$ and $1/12^2 = 1/144 \approx 0.007$. The decimal approximation with a two digit denominator is $14/10$, with accuracy 0.014. The continued fraction approximant is usually much better than the decimal approximation.

Computing Approximations

When n is large it is difficult to compute the rational number form of an nth-order approximant by hand. We can continue our computations with the CAS. Suppose we wish to find a best rational approximation to an irrational number a with accuracy of at least ACC.

First we need a function to generate continued fraction representations.

```
> # Continued fraction representation of r to n terms.
> cfr:= proc(r,n)
≫        local i,spec,x;
≫        x:= r;
≫        for i from 1 to n do
≫                spec[i] := trunc(x);
≫                x:= cfv(x);
≫                # Check if cf repn.  is done.
≫                if x = 1 then break fi;
≫        od;
≫        RETURN(spec);
≫ end;
```

Now we need a function to convert a continued fraction representation back into a decimal.

```
> # Convert cf repn to numeric form.
> decf:= proc(a)
≫        local i,val,n;
≫        # Find last number stored in a:
≫        n:= 1;
≫        while type(a[n],integer) do n:= n+1 od;
≫        # Compute cf value:
≫        val:= a[n-1];
```

```
≫          for i from 1 to n-2 do
≫                  val:= a[n-1-i] + 1/val;
≫          od;
≫          RETURN(val);
≫ end;
```

We use the above two functions to convert x to a continued fraction representation that is accurate to within ACC.

```
> # cfr of x accurate to within ACC.
> acfr:= proc(x,ACC)
≫          local spec,n;
≫          # A starting value.
≫          n:= 1;
≫          spec[n]:= trunc(x);
≫          # Test and generate while not accurate.
≫          while abs(evalf(x - decf(spec))) > ACC do
≫                  n:= n+1;
≫                  spec:= cfr(x,n);
≫          od;
≫          RETURN(spec);
≫ end;
```

Finally, we include a procedure to print out the spectrum that has been stored in spec.

```
> # Print spectrum held in spec[1],spec[2],...
> pspec:= proc(spec)
≫          local n;
≫          n:= 1;
≫          while type(spec[n],integer) do
≫                  print(spec[n]);
≫                  n:= n+1;
≫          od;
≫ end;
```

Exercises

1. Some CASs have built-in functions for partial fraction operations. Check your CAS for these functions, and check our computations.

2. Find the continued fraction representation of $23/5$.

3. Find the first 20 elements in the spectrum of $\sqrt{3}$.

4. Find the number with spectrum $[1,1,1,1]$.

5. Find an irrational number with an irregular spectrum.

6. Find a famous (irrational) mathematical constant with a regular spectrum.

7. Find a famous (irrational) mathematical constant with an irregular spectrum.

8. Find a best rational approximation to $\sqrt{3}$ that is accurate to within 10 decimal digits. How much better is it than the decimal approximation with the same number of digits in the denominator?

9. Find a best rational approximation to $\sqrt{7}$ that has a denominator of at most 1000. It should be as close to $\sqrt{7}$ as possible.

10. f has discontinuities at integer values. f^2, defined by $f^2(x) = f(f(x))$ has a further set of discontinuities within each interval $(n, n+1)$. This makes it very difficult to graph. Try out your graphing system on this function.

11. Show that $\lim_{n \to \infty} (\tau_n - n) = 0$ analytically.

Chapter 4

Integration

4.1　Approximating Area

Prerequisites

- The Mean Value Theorem

Discussion

Area formulas exist for a few types of planar regions such as rectangles, triangles, trapezoids, circles, etc. However, no area formulas exist for the large majority of planar regions. For example, what is the surface area of Lake Erie? How do you determine the surface area of an oil slick? When we cannot determine area exactly, we rely on approximations. As always when working with an approximation, determining an error bound is extremely important, that is, determining how good the approximation is.

The standard way to proceed in approximating the area of a planar region is to divide the region into small pieces, approximate each small piece with a geometrical shape region whose area is known (e.g., a rectangle), and then sum up the areas of the approximating regions. The accuracy of the approximation can usually be increased by dividing the region into smaller pieces. An error bound is obtained by forming two approximations, one that is too small and one that is too large. The absolute value of the difference of these two approximations is then an error bound for either of the approximations.

In order to develop an understanding of the process of approximating the area of a region to within a prescribed accuracy (i.e., error bound), we begin by considering the area of the planar region bordered below by the x-axis over the interval [a,b], above by the graph of a continuous function $y = f(x)$, on the left by the vertical line $x = a$, and on the right by the vertical line $x = b$. This region is often referred to as the *area under the graph* $y = f(x)$ *over the interval* $[a, b]$. The region is divided into vertical strips (i.e., small pieces) by *partitioning* the interval into n subintervals of equal length $(\frac{b-a}{n})$. A partition P_n is denoted by listing the set of points that determine the n subintervals: $P_n = \{a = x_0, x_1, x_2, \ldots, x_n = b\}$. The next step is to approximate each vertical strip by a rectangle. A *lower* approximation is formed by choosing the height of the rectangle to be the minimum value of the function over the subinterval. Similarly, an *upper* approximation is formed by choosing the height of the rectangle to be the maximum value of the function over the subinterval. The sum of the lower approximations is called the *lower sum* determined by f and P and is denoted by $L(f, P_n)$. Similarly, the sum of the upper approximations is called the *upper sum* and is denoted by $U(f, P_n)$.

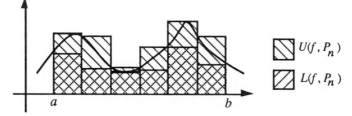

We denote the minimum and maximum values of the function f over the ith subinterval by $\min_i f$ and $\max_i f$, respectively. The lower and upper sums are now expressed in summation notation by

$$L(f, P_n) = \sum_{i=1}^{n} \min_i f \; \frac{b-a}{n}$$

$$= \min_1 f \; \frac{b-a}{n} + \min_2 f \; \frac{b-a}{n} + \cdots + \min_n f \; \frac{b-a}{n}$$

$$U(f, P_n) = \sum_{i=1}^{n} \max_i f \; \frac{b-a}{n}$$

$$= \max_1 f \; \frac{b-a}{n} + \max_2 f \; \frac{b-a}{n} + \cdots + \max_n f \; \frac{b-a}{n}$$

The area of a region bounded above by the graph of the continuous function f, below by the interval $[a, b]$, on the left by the line $x = a$, and on the right by the line $x = b$, is *sandwiched* between the monotonically increasing sequence, $\{L(f, P_n)\}$, of lower sums (approximations) and the monotonically decreasing sequence, $\{U(f, P_n)\}$, of upper sums (approximations).

There exists an easily derived error bound expression when the bounding function in question (i.e., the function whose graph bounds the region from above) is differentiable and has a bounded derivative. Assume $|f'(x)| \leq B$ for all x in [a,b]. The key step in the derivation is an application of the Mean Value Theorem. (Note that $\max_i f - \min_i f \leq f(x_{i+1}) - f(x_i) = f'(c_i)(x_{i+1} - x_i) =$

$f'(c_i)\dfrac{b-a}{n}$ for some c_i in $[x_{i-1}, x_i]$. Thus $\max_i f - \min_i f \leq B\dfrac{b-a}{n}$.)

$$
\begin{aligned}
U(f, P_n) - L(f, P_n) &= \sum_{i=1}^{n}[\max_i f - \min_i f]\,\frac{b-a}{n} \\[2mm]
&\leq \sum_{i=1}^{n}\left[B\frac{b-a}{n}\right]\frac{b-a}{n} \qquad (\text{ by the Mean Value Th.}) \\[2mm]
&= \sum_{i=1}^{n} B\left(\frac{b-a}{n}\right)^2 \\[2mm]
&= nB\left(\frac{b-a}{n}\right)^2 \qquad \left(\text{since } \sum_{i=1}^{n} B = nB\right) \\[2mm]
&= B\frac{(b-a)^2}{n}
\end{aligned}
$$

A major difficulty in computing $L(f, P_n)$ and $U(f, P_n)$ is determining the minimum and maximum values of f over each subinterval. This difficulty can be avoided by realizing that since $\min f \leq f(x) \leq \max f$ over any subinterval, all approximations of the form $S(f, P_n) = \sum_{i=1}^{n} f(x_i)\dfrac{b-a}{n}$, where x_i is any point in the ith subinterval, lie between $L(f, P_n)$ and $U(f, P_n)$. That is, $L(f, P_n) \leq S(f, P_n) \leq U(f, P_n)$. Furthermore since $B\dfrac{(b-a)^2}{n}$ is an error bound for both $L(f, P_n)$ and $U(f, P_n)$, $B\dfrac{(b-a)^2}{n}$ is an error bound for all approximations $S(f, P_n)$.

For a continuous function f, all sequences of sums of the form $S(f, P_n) = \sum_{i=1}^{n} f(x_i)\dfrac{b-a}{n}$ converge to a common limit. (Note that this includes $L(f, P_n)$ and $U(f, P_n)$.) This result is easily established in the case where f has a bounded derivative, since

$$
U(f, P_n) - L(f, P_n) \leq B\frac{(b-a)^2}{n}
$$

The area of the type of region being considered is defined to be the common limit of the sequences of approximating sums.

A typical type of area approximation problem is to ask for an approximation of the area of a region within a stated accuracy (i.e., error bound). We will illustrate the process by approximating the area of a trapezoid, a region for which there exists an area formula. This will enable us to numerically check our answer.

Example 1

We will approximate the area under the graph of $f(x) = 2x$ over the interval $[2, 4]$ within an accuracy of 0.01. The region is a trapezoid with area 12. (The area of a trapezoid is one-half the sum of the heights times the base.)

We begin by using the error bound expression $B\dfrac{(b-a)^2}{n}$, where $|f'(x)| \leq B$ on $[2, 4]$, to determine n, the desired number of subintervals in the partition of $[2, 4]$. Since $f'(x) = 2$, we set $B = 2$. Then $B\dfrac{(b-a)^2}{n} = 2\dfrac{(4-2)^2}{n} = \dfrac{8}{n}$. Now setting the error bound expression equal to the prescribed accuracy, $\dfrac{8}{n} = 0.01$, and solving for n gives $n = 800$.

We now compute the *right-hand* sum (i.e., select x_i in the summation expression to be the right-hand endpoint of the ith subinterval). Thus we set $x_i = 2 + i\frac{4-2}{800} = 2 + i\frac{2}{800}$. We want to determine the sum:

$$\sum_{i=1}^{800} f(2 + i\frac{2}{800})\, \frac{2}{800}$$

To do this, we use our CAS. First we define our function f and then use the sum command.

```
> f:= proc(x) 2*x end;
> sum(f(2+i*2/800)*2/800, i=1..800);
```

returns $2401/200 = 12.005$. Note that our approximation of 12.005 is within 0.01 of 12, which is the exact area.   △

Often the easiest way to determine a bound for the derivative is to plot the absolute value of the derivative over the domain of the the integral and then read off an upperbound from the plot. (Why plot the absolute value of the derivative rather than just the derivative?)

Example 2

We will approximate the area under the graph of $f(x) = \sqrt{1 + x^3}$ over the interval $[4, 6]$ within an accuracy of 0.1 using a *left-hand* sum.

We will use the error bound expression $B\dfrac{(b-a)^2}{n}$, where $|f'(x)| \leq B$ on $[4, 6]$, to determine n, the desired number of subintervals in the partition of $[4, 6]$. To determine a suitable value for B, we define our function and then plot its derivative over $[4,6]$.

```
> f:= proc(x) sqrt(1+x^3) end;
> plot(diff(abs(f(x)),x), x= 4..6);
```

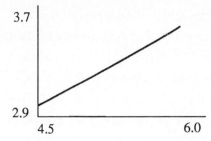

Thus $|f'(x)| \leq 3.7$ over $[4,6]$. Substituting $B = 3.7$ into the error bound expression and setting the expression equal to the accuracy of 0.1 yields

$$B\frac{(b-a)^2}{n} = 3.7\frac{(6-4)^2}{n} = \frac{3.7(4)}{n} = \frac{1}{10}$$

Solving for n gives, $n = 148$.

We now compute the *left-hand* sum (i.e., select x_i in the summation expression to be the left-hand endpoint of the ith subinterval). Thus we set $x_i = 4 + i\frac{4-2}{148} = 4 + i\frac{2}{148}$ where i goes from 0 to 147. We want to determine the sum:

$$\sum_{i=0}^{147} f(4 + i\frac{2}{148}) \frac{2}{148}$$

To do this, we use our CAS.

```
> sum(f(4+i*2/148)*2/148, i=0..147);
```

The answer is given in screenfulls of radical expressions since Maple computes in exact arithmetic and does not approximate radicals in decimal form unless instructed to do so. Thus, to get a decimal answer, we evaluate the previous expression to decimal form.

```
> evalf(");
```

obtaining 22.51914562 or 22.5 with accuracy of 0.1.   △

Exercises

1. Approximate the area under the graph of $f(x) = \sin(x)$ over the interval $[0, \pi]$ with an error bound of 0.1 or less. First determine n, the number of subintervals required to guarantee the desired accuracy and then compute the value of the approximation. (Hint: The exact area is 2.)

   (a) Let x_i be the right endpoint of each subinterval.

(b) Let x_i be the left endpoint of each subinterval.

(c) Let x_i be the midpoint of each subinterval.

(d) Compare the number n of subintervals for each of the three choices of x_i. Which choice is best?

2. Repeat Exercise 1 using the function $f(x) = \sqrt{1 + x^4}$ over the interval $[2, 5]$.

3. Repeat Exercise 1 using the function $f(x) = x \sin(x) + 2$ over the interval $[0, 4]$.

4. Repeat Exercise 1 using the function $f(x) = \sqrt[3]{1 + x^2}$ over the interval $[0, 4]$.

5. The number n_0 of subintervals given by the error estimate is a conservative estimate. If you use n_0 subintervals, you are guaranteed the desired accuracy. This value of n_0 depends only on a bound for the derivative; thus only some of the information about the function f is used. It may be possible to obtain the desired accuracy with fewer subintervals. Consider the region under the graph of $f(x) = x^2$ over the interval $[0, 3]$. The exact area is 9.

(a) Determine the number of subintervals, n_0, from the error estimate that guarantees an accuracy of 0.1.

(b) (left-hand sum) Let x_i be the left-hand end point in the ith subinterval. How many subintervals are *actually* required to obtain the desired accuracy?

(c) (middle sum) Let x_i be the mid-point in the ith subinterval. How many subintervals are *actually* required to obtain the desired accuracy?

(d) (right-hand sum) Let x_i be the right-hand endpoint in the ith subinterval. How many subintervals are *actually* required to obtain the desired accuracy?

6. Repeat Exercise 5 for the function $f(x) = \sin(1/x)$ on the interval $[0.4, 1]$. Plot f over the interval, to see what it looks like.

7. Extend the development given in this section for approximating the area of a region. Give a worked example for each of the 3 extensions requested.

(a) Replace the restriction that the lower boundary of the region be an interval $[a, b]$ with the restriction that the lower boundary be the graph of a positive function over an interval $[a, b]$.

(b) Remove the restriction that the region lie on or above the x-axis.

(c) Remove the (implied) restriction that the upper or lower boundary of the region is the graph of a single function.

8. Approximate the area of the region bounded by the graphs of $f(x) = \sin(x)$ and $g(x) = x^2$ over the interval $[-1, 1]$ with an error bound of .01 or less.

9. Approximate the area of the region bounded by the graphs of $f(x) = \sin(x)$ and $g(x) = \cos(x)$ over the interval $[-1, 8]$ with an error bound of .01 or less.

10. Approximate the area of the region bounded by the graphs of $f(x) = x^2$, $g(x) = \dfrac{4}{x}$, and $h(x) = \dfrac{3}{2}x + \dfrac{5}{2}$ for $-1 \le x \le 2$ with an error bound of .01 or less.

4.2   Numerical Integration

Prerequisites

- An introduction to the definite integral

- The section on Approximating Area

Discussion

Polynomials constitute the only large and important class of functions whose integrals can be evaluated exactly. There are a few special radical expressions, certain trigonometric expressions, and "nice" rational expressions (i.e., the denominators can be factored) as well as a few exponential or logarithmic forms that can be evaluated. However for evaluation purposes, the vast majority of integrals encountered in applied or theoretical settings are approximated using a numerical integration technique. We will work with three numerical integration methods:

Riemann Sum   Trapezoidal Rule   Simpson's Rule

The procedure in each method is similar; that is, partition the interval $[a, b]$ into subintervals of equal length and then approximate the graph of f over each subinterval with a "simpler" function.

For Riemann Sums, f is approximated over each subinterval by a constant function.

For the Trapezoidal Rule, f is approximated over each subinterval by a linear function.

For Simpson's Rule, f is approximated over pairs of subintervals by a quadratic function.

Geometrically the result of the partitioning $[a, b]$ and approximating the function is to subdivide the region between the graph of f and the x-axis into subregions whose areas can be computed.

For Riemann Sums, the subregions are rectangles:

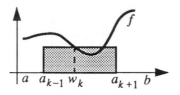

For the Trapezoidal Rule, the subregions are trapezoids:

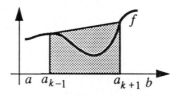

For Simpson's Rule, the subregions are parabolic:

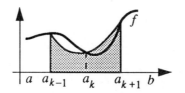

The numerical approximations are then computed by summing up the areas of the subregions. When the function f is continuous, each of the methods gives an approximation satisfying a prescribed accuracy by partitioning $[a, b]$ into a sufficiently large number of subintervals. The partition $\{a = x_0, x_1, ..., x_{n-1}, x_n = b\}$ subdivides the interval $[a, b]$ into n subintervals, each of length $\frac{b-a}{n}$. The expressions for approximating $\int_a^b f(x)dx$ are:

Riemann Sum

$$\int_a^b f(x)dx \approx [f(w_0) + f(w_1) + f(w_2) + \cdots + f(w_n)]\frac{b-a}{n}$$

$$= \sum_{k=1}^n f(w_k)\frac{b-a}{n} \quad \text{where } w_k \text{ is a point in } [x_{k-1}, x_k]$$

Trapezoidal Rule

$$\int_a^b f(x)dx \approx \frac{1}{2}[f(x_0) + 2f(x_1) + 2f(x_2) + \cdots + 2f(x_{n-1}) + f(x_n)]\frac{b-a}{n}$$

$$= \ [f(x_0) + f(x_n) + \sum_{k=1}^{n-1} 2f(x_k)] \ \frac{b-a}{2n}$$

Simpson's Rule

$$\int_a^b f(x)dx \ \approx \ \frac{1}{3}[f(x_0) + 4f(x_1) + 2f(x_2) + 4f(x_3)$$

$$+ \cdots + 2f(x_{2n-2}) + 4f(x_{2n-1}) + f(x_{2n})] \ \frac{b-a}{2n}$$

$$= \ [f(x_0) + f(x_n) + \sum_{k=1}^{n/2} 4f(x_{2k-1}) + \sum_{k=1}^{(n-1)/2} 2f(x_{2k})]$$

In general, the accuracy of the approximations is improved by increasing the value of n (the number of subintervals). Note that for Simpson's Rule, n must be an even integer since the subregions are formed over pairs of subintervals. As with any approximation method, it is important to analyze the error. That is, determine an error bound. Error bounds that depend on the derivatives of f and n exist for the three approximation methods. They are:

Error Bound for the Riemann Sum Approximation

$$| \text{ Riemann Sum Approx.} - \int_a^b f(x)dx \ | \leq \frac{B(b-a)^2}{n}, \qquad \text{where } |f'(x)| \leq B$$

Error Bound for the Trapezoidal Rule Approximation

$$| \text{ Trapezoidal Rule Approx.} - \int_a^b f(x)dx \ | \leq \frac{B(b-a)^3}{12n^2}, \qquad \text{where } |f''(x)| \leq B$$

Error bound for Simpson's Rule Approximation

$$| \text{ Simpson's Rule Approx.} - \int_a^b f(x)dx \ | \leq \frac{B(b-a)^5}{180n^4}, \qquad \text{where } |f^{(4)}(x)| \leq B$$

Note that although the same letter, B, is used to denote a bound in each of the three error bound formulas, the expressions being bounded are different. For Riemann Sums, B is a bound on the *first* derivative over the interval $[a, b]$; for the Trapezoidal Rule, B is a bound on the *second* derivative over the interval; and for Simpson's Rule, B is a bound on the *fourth* derivative over the interval.

We call attention to the fact that when the number of subintervals is doubled, the accuracy increases by at least a factor of 2 for the Riemann Sum method, 4 for the Trapezoidal Rule, and 16 for Simpson's Rule. (See exercise 3.)

The typical type of question with which we will be concerned is to approximate the value of an integral within a prescribed accuracy. There are three steps in such a problem: the first is to bound the first, second, or fourth derivative (depending on the approximation method); the second is to set the error bound expression equal to the accuracy and solve for n; the third is to compute the summation expression. These computations are easily accomplished using the differentiation, plot, solve, and sum commands of a CAS.

Most CASs have some built-in numerical integration method. For students whose CAS does not have a Riemann Sum, Trapezoidal Rule, or Simpson's Rule command, we include at the end of this section user-defined procedures for these commands.

Example 1

We approximate $\int_1^4 \sin(x^2)dx$ with an accuracy of at least 10^{-4} using Simpson's Rule. (The procedures for applying the Riemann Sum or Trapezoidal Rule are similar.)

We first determine a bound for the fourth derivative of $\sin(x^2)$ using the CAS differentiate and plot commands.

```
> f:= proc(x) sin(x^2) end;
> d4f:= diff(f(x),x$4);
> plot(d4f, x=1..4);
```

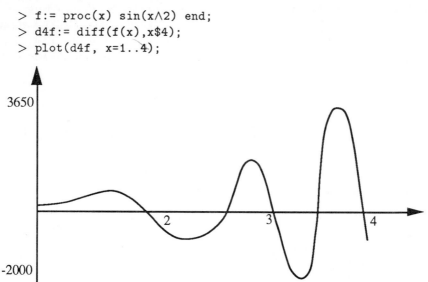

The plot shows that $B = 3650$ is a bound for the fourth derivative of $\sin(x^2)$ over [1,4].

We now determine the value of n (number of subintervals) by setting the error bound expression equal to the prescribed accuracy and solving for n.

$$\text{Simpson's Rule Error Bound} = \frac{B(4-1)^5}{180n^4} = \frac{3650(3^5)}{180n^4} \le 10^{-4}$$

```
> n:= solve((3650*3^5)/(180*n^4)= 10^(-4),n);
```

this gives one positive, one negative, and two imaginary solutions. The positive solution rounded up to the nearest even integer is $n = 84$.

The final step is to evaluate the Simpson's Rule approximation to the integral of f over $[1, 4]$ with $n = 84$.

```
> approximation:= simpson(f,1,4,84);
```

$$0.4369$$

to four significant digits.

The corresponding B and n values for the Riemann Sum approximation and for the Trapezoidal Rule approximation are:

Riemann Sum: $B = 9$ $n = 120,000$

Trapezoidal Rule: $B = 62$ $n = 682$ $\triangle$

Using our CAS to obtain a multiplot of the error bound expression and the prescribed accuracy provides an alternative to using a solve command for determining an appropriate value for the number of subintervals. Since the error bound is a decreasing function of n, we want to determine the value of n where the graph of the error bound function crosses the line representing the accuracy. For the above example, the multiplot would be

Note that the number of subintervals used, 84, is based not on the specific function, $f(x) = \sin(x^2)$, but on the bound for the fourth derivative. Eighty four subintervals are guaranteed to work for any function with the same bound. However, for this particular function, if we use 42 subintervals, then the value returned is 0.436943, which is 0.4369 using four significant digits.

Error bounds are usually very conservative since they must hold for *worst case* scenarios. Often a considerably smaller value of n will yield the desired result. A method for finding a *heuristic* value of n is to experiment with different values of n until a number n^* is obtained such that the answers found from using n^* and a larger n, say $2n^*$, differ by (substantially) less than the prescribed accuracy. We can use our CAS and a **for** loop to compute a sequence of approximation values. For example, the following **for** loop computes the Simpson's Rule approximations in the previous problem for even values of n from 20 to 80.

```
> for n from 10 to 40 do
>>        print(2*n,simpson(sin(x^2), x=1..4,2*n)
>> od;
```

Example 2

We approximate $\displaystyle\int_{1}^{4}\frac{\sqrt{1+x^3}}{x}dx$ with an accuracy of at least 10^{-4} using a heuristic value for n, the number of subintervals. We use a **for** loop to generate the numerical results using Simpson's Rule for n ranging from 10 to 20 in steps of 2. (Note that Simpson's Rule requires n to be an even integer.)

```
> for n from 5 to 10 do
≫         print(2*n,simpson(sqrt(1+x∧3)/x, x=1..4,2*n)
≫ od;
```

```
10,   4.936762443
12,   4.936629086
14,   4.936568936
16,   4.936538750
18,   4.936538750
20,   4.936512763
```

Since the first four decimal places do not change for $n = 14$ through 20, the desired accuracy of 10^{-4} seems to have been obtained with $n = 14$. We now double this value and check the approximation with $n = 28$.

```
> simpson(sqrt(1+x∧3)/x, x=1..4,28);
```

$$4.936497813$$

Since the approximations for $n = 14$ and 28 differ by less than .00007, we consider 4.9365 as an acceptable (heuristic) approximation.

The theoretical value of n is 26 (obtained by applying the error bound formula) which gives an approximation of 4.936500700.   △

An interesting area for exploration is the relationship between the robustness of an Error Bound and the integrand of an integral. For example, for a given number of subintervals is the error in the Trapezoidal approximation of $\int_{0}^{3} x^3\, dx$ less than or greater than the error in the Trapezoidal approximation of $\int_{0}^{3} x^5\, dx$? (Why?) Or, what type of function would provide a "worst case" for the Trapezoidal Rule? Or, is there a relationship between the concavity of the graph of f and its Riemann Sum Error Bound? Or, for what type of function is the Trapezoidal Rule a more efficient approximation method than is Simpson's Rule? Or, does the error bound formula for Simpson's Rule imply that any polynomial of degree ≤ 3 can be integrated exactly using Simpson's Rule?

Riemann Sum Procedure

This procedure evaluates at the midpoint of each subinterval. This can be changed by changing **evalpt** below. Note that **f** must be a function, not an expression.

```
> riemann:= proc(f,left,right,npan)
> # Riemann Sum.
> # Input:
> # f = function to be integrated.
> # left = left endpoint of interval.
> # right = right endpoint of interval.
> # npan = number of subintervals.
>        # variables local to this procedure.
>        local sum, evalpt, width;
>        # Initialize variables.
>        sum:= 0;
>        width:= (right-left)/npan;
>        # Evaluate at midpoints.
>        evalpt:= left + width/2;
>        for i from 1 to npan do
>                sum:= sum + f(evalpt)*width;
>                evalpt:= evalpt + width;
>        od;
>        RETURN(evalf(sum));
> end;
```

Trapezoidal Rule Procedure

Note that **f** must be a function, not an expression.

```
> trapezoid:= proc(f,left,right,npan)
> # Trapezoid Rule.
> # Input:
> # f = function to integrate.
> # left = left endpoint of interval.
> # right = right endpoint of interval.
> # npan = number of subintervals.
>        # Variables local to this procedure.
>        local sum,evalpt,width,i;
>        # Initialize variables.
>        sum:= 0;
>        width:= (right-left)/npan;
>        evalpt:= left;
>        for i from 0 to npan do
>                sum:= sum + f(evalpt);
>                evalpt:= evalpt + width;
>        od;
>        sum:= 2*sum-f(left)-f(right);
>        sum := sum*width/2;
```

```
>>          RETURN(evalf(sum));
>> end;
```

Simpson's Rule Procedure

Note that **f** must be a function, not an expression.

```
> simpson:= proc(f,left,right,npan)
>> # Simpson's Rule.
>> # Input:
>> # f = function to be integrated.
>> # left = left endpoint of interval.
>> # right = right endpoint of interval.
>> # npan = number of subintervals (must be even).
>>        # Variables local to this procedure.
>>        local i,sumodd,sumeven,halfnpan,evalpteven,
>>                  evalptodd,width,width2;
>>        # Initialize variables.
>>        halfnpan:= npan/2;
>>        width:= (right - left)/npan;
>>        width2:= 2*width;
>>        sumodd:= 0;
>>        sumeven:= 0;
>>        evalpteven:= left + width2;
>>        evalptodd:= left + width;
>>        # Compute Simpson sum.
>>        for i from 1 to halfnpan do
>>             sumeven := sumeven + f(evalpteven);
>>             evalpteven:= evalpteven + width2;
>>             sumodd:= sumodd + f(evalptodd);
>>             evalptodd:= evalptodd + width2;
>>        od;
>>        sumeven:= 2*sumeven + 4*sumodd;
>>        sumeven:= sumeven + f(left) - f(right);
>>        sumeven:= sumeven*width/3;
>>        RETURN(evalf(sumeven));
>> end:
```

Exercises

1. Fill in the following table for $\int_{-1}^{2} x^5 dx = 10.5$

n	Approx. Value	(Actual) Error	Error Bound
10			
20			
30			
50			
80			
100			

using:

(a) Riemann Sum, (b) Trapezoidal Rule, and (c) Simpson's Rule.

What observations can you make (e.g., how does the actual error compare with the error bound)?

2. Fill in the table for $\int_0^3 \sin(\sin(x))dx$, $ACC = 0.01$.

Method	n	Approx. Value	Error Bound	n^* (heuristic)
Riemann Sum				
Trapezoidal Rule				
Simpson's Rule				

3. (a) Fill in the table of errors for approximating $\int_0^4 \sqrt{1 + x^2}dx = 9.293567$ using the Riemann Sum, Trapezoidal Rule, and Simpson's Rule for values of n from 2 to 20 in steps of 2.

n	Error(Riemann Sum)	Error(Trapezoidal)	Error(Simpson)
2			
4			
8			
—	——	——	——
—	——	——	——
20			

Hint: Error(Simpson) = 9.293567 $-$ simpson(f(x),x=0..4,2*n).

(b) Verify the fact that when the number of subintervals is doubled, the accuracy increases by at least a factor of 2 for the Riemann Sum method, 4 for the Trapezoidal Rule, and 16 for Simpson's Rule.

In Exercises 4-8, give a complete report: the approximation method, the value of the bound B, the theoretically estimated number of subintervals required, the approximate integral value, and a heuristic estimate of the minimum number of subintervals required for the given accuracy.

4. $f(x) = \dfrac{1}{1 + x^3}$, interval $[1, 2]$, error 0.01

5. $f(x) = \sqrt{1 - x^3}$, interval $[0,1]$, error 0.01

6. $f(x) = \sin(\sin(x))$, interval $[0,1]$, error 0.01

7. $f(x) = \sin(x^2 + 3x + 1)$, interval $[1,2]$, error 0.01

8. $f(x) = \dfrac{1}{\sqrt{1 + x^3}}$, interval $[1,2]$, error 0.01

9. Consider the functions $f(x) = x^2$ and $g(x) = x^4$. Use a **for** loop to form a table as n goes from 1 to 10 of the actual errors when $\int_{-2}^{2} f(x)\, dx$ and $\int_{-2}^{2} g(x)\, dx$ are approximated using the Trapezoidal Rule. (The actual error is the value of the integral minus the numerical approximation.) Make a conjecture, based on your table of errors, as to which integral is better approximated using the Trapezoidal Rule for a given value of n. Justify your conjecture on the basis of the curvatures of the functions f and g. Explore the relationship between the curvature of the graph of a function and the accuracy of the Trapezoidal approximation to its integral for a given value of n.

10. Repeat exercise 9 using the functions $f(x) = x^4$ and $g(x) = x^6$, Simpson's Rule, and steps of $2n$ as n goes from 1 to 10.

11. For each of the three numerical integration methods, find an example of an integral where the theoretical number of subintervals needed was much, much more than the "heuristic number." (You need to set the accuracy value and also quantify your interpretation of "much, much more.") Hint: Compare how the heuristic and theoretical number of subintervals increase when approximating $\int_0^4 x^k\, dx$ as k increases, say $k = 4, 5, 7$, etc.. Also consider the role of the length of the interval when computing the theoretical error bound.

12. For each of the three numerical integration methods, find an example of an integral where the theoretical number of subintervals needed was about the same as the "heuristic number." Hint: Consider approximating $\int_0^3 e^x\, dx$.

13. Find an integral where the approximation obtained using the heuristic number of subintervals was very, very wrong. (You need to quantify your interpretation of "very, very wrong.") Hint: Consider a function whose graph has a "spike," say $f(x) = 1$ for $0 \le x \le 1.99$ and $f(x) = 1 + 100x$ for $1.99 < x \le 2$.

4.3   Integration by Substitution

Prerequisites

- The chain rule for differentiation of compositions of functions

Discussion

Most of the easier techniques of finding antiderivatives (or integration) involve using a differentiation rule "backwards." For the case of the integral of a sum, this is straight-forward and easily recognized; in the case of the chain rule (whose corresponding integration technique is called "substitution"), the pattern is not so easily recognized. The exercises in this section are designed to give you some practice in recognizing the form of a function whose antiderivative can be found by the method of substitution.

Recall that the chain rule tells us $\dfrac{d}{dx}f(g(x)) = f'(g(x))g'(x)$. Thus a function of the form $f'(g(x))g'(x)$ has $f(g(x))$ as its antiderivative. Using integral notation; $\int f'(g(x))g'(x)dx = f(g(x)) + const$. Some familiar manifestations of this pattern are:

$$\int 2xe^{x^2}\,dx \;=\; e^{x^2} + const$$

$$\int (2x+3)(x^2+3x-7)^5\,dx \;=\; \frac{1}{6}(x^2+3x-7)^6 + const$$

$$\int 3x^2\cos x^3\,dx \;=\; \sin x^3 + const$$

Exercises

In each of the following exercises you will be given several integrals. Use the CAS to try to evaluate them and by observing the results, determine which one could be done by the method of substitution discussed above.

1. (a) $\int \sqrt{x^2+2}\,dx$ (b) $\int x\sqrt{x^2+2}\,dx$

   (c) $\int x^2\sqrt{x^2+2}\,dx$ (d) $\int x^3\sqrt{x^2+2}\,dx$

2. (a) $\displaystyle\int \frac{dx}{\sqrt{x^2-5}}$ (b) $\displaystyle\int \frac{x\,dx}{\sqrt{x^2-5}}$

   (c) $\displaystyle\int \frac{x^2\,dx}{\sqrt{x^2-5}}$ (d) $\displaystyle\int \frac{x^3\,dx}{\sqrt{x^2-5}}$

3. (a) $\int \sin(x^2)dx$ (b) $\int x\sin(x^2)dx$

   (c) $\int x^2\sin(x^2)dx$ (d) $\int x^3\sin(x^2)dx$

4. (a) $\displaystyle\int \sin\!\left(\frac{1}{x}\right)dx$ (b) $\displaystyle\int x\sin\!\left(\frac{1}{x}\right)dx$

(c) $\int x^{-1} \sin(\frac{1}{x}) \, dx$ (d) $\int x^{-2} \sin(\frac{1}{x}) \, dx$

4.4   Integration Formulas

Prerequisites

- Familiarity with some integration techniques, especially integration by parts

Discussion

The object of this section is to give the reader practice in working with indefinite integrals. We will develop some integration formulas found in integration tables using the CAS to carry out the computations.

Example

We will find a formula for $\int x^n \log(x) \, dx$. (Note that "log" denotes the natural logarithm.) If we try to integrate directly,

```
> int(x^n*log(x),x);
```

our CAS will probably fail since n is not assigned a value. However, we can combine the computational power of our CAS with our powers of observation to obtain a formula. We will try integrating $\int x^n \log(x) \, dx$ for several values of $n > 0$ and try to see a pattern:

```
> int(x*log(x),x);
```
   returns $\dfrac{x^2 \log(x)}{2} - \dfrac{x^2}{4}$

```
> int(x^2*log(x),x);
```
   returns $\dfrac{x^3 \log(x)}{3} - \dfrac{x^3}{9}$

```
> int(x^3*log(x),x);
```
   returns $\dfrac{x^4 \log(x)}{4} - \dfrac{x^4}{16}$

By now we see a pattern and conjecture that

$$\int x^n \log(x) \, dx = \frac{x^{n+1} \log(x)}{n+1} - \frac{x^{n+1}}{(n+1)^2} + C$$

for $n > 0$. Differentiation is much easier for computers as well as for people, and our CAS should be able to verify our conjecture:

```
> diff(x^(n+1)*log(x)/(n+1)-x^(n+1)/(n+1)^2,x);
```

returns

$$x^n \log(x)$$

Thus our conjecture was correct. △

The student should realize that integrals of functions such as $x^n \log(x)$, $\sin^n(x)$,

and $x^n e^x$ can be solved recursively using integration by parts. This is the integration technique that some CASs use for functions of these types.

Exercises

1. Develop an integration formula for $\int x \log(x^n) \, dx$.

   (a) Using your CAS, define the function $f(x, n) = x \log(x^n)$, n a positive integer:

   $$> \texttt{f := proc(x,n) x*log(x\char`\^n) end;}$$

   Then compute $\int f(x, n) \, dx$ for $n = 1, 2, 3, 4, \ldots$ Continue until you are able to conjecture a formula.

   (b) Check your conjecture by computing $\int x \log(x^n) \, dx$ using integration by parts. Hint: Note $\log(x^n) = n \log(x)$.

   (c) Check your conjecture using the CAS.

2. Find a formula for $\int x^n e^x \, dx$. Follow the steps in Exercise 1.

3. Find a formula for $\int x^n \sin(x) \, dx$.

4. Find a formula for $\int \sin^n(x) \, dx$. Follow the steps in Exercise 1. Hint: Look for a relation between the integrals for $f(x, n)$ and $f(x, n-2)$.

5. Find a formula for $\int \cos^n(x) \, dx$. Follow the steps in Exercise 1. Hint: Look for a relation between the integrals for $f(x, n)$ and $f(x, n-2)$.

4.5   Functions Defined by Definite Integrals

Prerequisites

- The Section on Numerical integration (subsection on Riemann Sums and Simpson's Rule)

- The section on Extremal Values

Discussion

Since an integral of the form $\int_c^x f(t)\,dt$ defines a function of x, we may analyze it as we would any other function; that is, determine if the function is continuous, differentiable, integrable; determine where the function is monotone; determine the extremal values; determine the graphical properties – concavity, intercepts, asymptotes; etc. If we can find a simple anti-derivative, then we say that we can evaluate the integral in *closed form*. Analyzing an integral that cannot be evaluated in closed form provides a richer and more challenging experience than is the case when the integral can be evaluated in closed form. A good practice procedure is to analyze a known integral (one that can be evaluated in closed form) without first integrating and then using the integrated form as a "check."

Example

As an example we will consider a case where the integration can be carried out (and a closed form expression for f found). It is usually a good idea to try a new technique on a problem where the answer can be checked by other methods. Consider the function f defined by $f(x) = \int_2^x (t^2 + 2)\,dt$, $-3 \le x \le 5$. First we sketch the graph of $y = f(x)$ without evaluating the integral. Then we use the evaluated form of the integral to check the sketch.

By inspection, we see that (i) f is differentiable (from the Fundamental Theorem of Calculus) and thus is continuous, (ii) $f(2) = 0$, and (iii) f is negative for $x < 2$ and f is positive for $x > 2$ since the integrand is positive. Applying the Fundamental Theorem of Calculus yields $f'(x) = x^2 + 2 > 0$. Thus f is monotone increasing over the interval $[-3, 5]$. Hence f has a minimum value at $x = -3$ and a maximum value at $x = 5$.

We will use Simpson's Rule with 50 subdivisions to evaluate $f(-3)$ and $f(5)$. Since the Simpson's Rule algorithm requires the lower limit of integration to be less than the upper limit, we must transform the integral expression for $f(-3)$ to interchange the limits of integration. That is

$$f(-3) = \int_2^{-3} (t^2 + 2)\,dt = -\int_{-3}^2 (t^2 + 2)\,dt$$

(Note that in the following CAS command, the simpson command is multiplied by -1.)

```
> -1*simpson(x^2+2,x=-3..2,50);        returns -21.6667
> simpson(x^2+2,x=2..5,50);            returns 45.000.
```

We determine the concavity by looking at the second derivative (the first derivative of the integrand):

```
> diff(x^2+2,x);
```

yields $2x$.

Thus the graph of f has negative concavity over $[-3,0)$, positive concavity over $(0,5]$, and an inflection point at $(0,0)$. To "tie down" our sketch, we use Simpson's Rule to evaluate f at a few points, say at $x = -1$ and 3. We can now sketch the graph of f over $[-3,5]$.

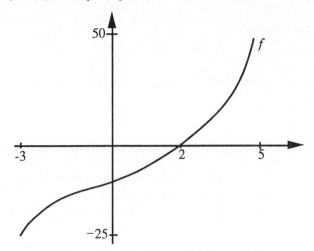

We now use our CAS to plot the integrated form of f:

```
> plot(int(f(t), t=2..x),x=-3..5);
```

(Note that we can compose CAS functions just as we compose any other functions.) This plot is similar to our hand sketch, thus verifying our method. △

Exercises

In Exercises 1 through 6, determine intercepts, concavity, inflection points, and extremal values, and then sketch the graph of $y = f(x)$.

1. $f(x) = \int_{-1}^{x} \sqrt{t^4 + 1}\, dt$

2. $f(x) = \int_{x}^{1} \sqrt{t^4 + 1}\, dt$

3. $f(x) = e^{-x} \int_{1}^{x} \log(t)\, dt, \quad x > 0$

4. $f(x) = 10 \int_{2}^{x} e^{-t} \log(t)\, dt, \quad x > 0$

5. $f(x) = \int_{1}^{x} \sqrt{t^3 + 1}\, dt, \quad x \geq -1$

6. $\int_{\pi}^{x} \sqrt{t} \sin(t)\, dt, \quad 0 \leq x \leq 4\pi$

4.6 Improper Integrals

Prerequisites

- Limits

- Integration techniques

- Improper integrals

- The sections on Limits of Functions and on the **for** loop for some parts

Discussion

If f is bounded on $[a, b]$ then we define $\int_a^b f(x)\, dx$ as a certain limit of Riemann sums (if this limit exists). Furthermore, the Fundamental Theorem of Calculus tells that under certain conditions, including the boundedness of f (and usually, in elementary texts, the continuousness of f on $[a, b]$), that $\int_a^b f(x)\, dx$ can be evaluated by finding an antiderivative of f and subtracting the antiderivative's value at a from its value at b. If f is unbounded on the interval $[a, b]$, we may or may not be able to make sense out of $\int_a^b f(x)\, dx$, and we certainly cannot use the fundamental theorem to evaluate this integral. We say that such an integral is *improper*. Recall that in the special case when f is continuous on $(a, b]$ but not at a we define

$$\int_a^b f(x)\, dx = \lim_{c \to a+} \int_c^b f(x)\, dx \qquad \text{(if it exists)}$$

If this limit exists, we say that the improper integral *converges*. This idea may be extended to each point where f is discontinuous (or undefined). We also extend the concept of definite integral to infinite intervals by defining (for example)

$$\int_a^\infty f(x)\, dx = \lim_{c \to \infty} \int_a^c f(x)\, dx \qquad \text{(if it exists)}$$

In this section we will use the CAS to help gain some experience in the recognition and evaluation of improper integrals.

Example

Consider the integral $\displaystyle\int_1^\infty \frac{1}{x^2}\, dx$. From the calculus of improper integrals, we know that this converges to 1. Let's see what happens if we look at some approximations.

Input	Output
> `int(1/x∧2,x=1..100);`	.9900000000
> `int(1/x∧2,x=1..200);`	.9950000000
> `int(1/x∧2,x=1..300);`	.9966666667

This seems to be approaching 1.

We could look at a large number of examples by automating the process – by using a **for** loop to vary the upper bound. The command is

```
> for k from 1 to 50 do
≫          print(evalf(int(1/x^2,x=1..(k*100))));
≫ od;
```

A final method is to define a function f whose value $f(x)$ is the integral $\int_1^x \frac{1}{t^2} \, dy$:

```
> f:= proc(x) evalf(int(1/t^2,t=1..x)) end;
```

Now we can apply the **limit** command:

```
> limit(f(x),x=infinity);
```

returns the answer of 1.   △

Exercises

Exercises 1 through 5 should be done in order. Exercises 1 through 3 provide background experience for 4, and 5 provides an application of 4.

1. Consider $\int_0^1 \frac{dx}{x^2}$. This clearly is an improper integral as $\frac{1}{x^2}$ is not defined

   and bounded throughout $[0, 1]$, although $\int_a^1 \frac{dx}{x^2}$ is defined for all $a \in (0, 1]$.

   Use the CAS to compute $\int_{0.1}^1 \frac{dx}{x^2}$, $\int_{0.001}^1 \frac{dx}{x^2}$, and $\int_{0.0001}^1 \frac{dx}{x^2}$. Do you think

   $\int_0^1 \frac{dx}{x^2}$ converges? If so, to what value?

2. Consider $\int_0^1 \frac{dx}{x}$. This also is an improper integral as $\frac{1}{x}$ is not defined and

   bounded throughout $[0, 1]$, although $\int_a^1 \frac{dx}{x}$ is defined for all $a \in (0, 1]$.

   Use the CAS to compute $\int_{0.1}^1 \frac{dx}{x}$, $\int_{0.001}^1 \frac{dx}{x}$, and $\int_{0.0001}^1 \frac{dx}{x}$. Do you think

   $\int_0^1 \frac{dx}{x}$ converges? If so, to what value?

3. Consider $\int_0^1 \frac{dx}{\sqrt{x}}$. This also is an improper integral as $\frac{1}{\sqrt{x}}$ is not defined

   and bounded throughout $[0, 1]$, although $\int_a^1 \frac{dx}{\sqrt{x}}$ is defined for all $a \in (0, 1]$.

Use the CAS to compute $\int_{0.1}^{1} \frac{dx}{\sqrt{x}}$, $\int_{0.001}^{1} \frac{dx}{\sqrt{x}}$, and $\int_{0.0001}^{1} \frac{dx}{\sqrt{x}}$. Do you think $\int_{0}^{1} \frac{dx}{\sqrt{x}}$ converges? If so, to what value?

4. Now consider $\int_{0}^{1} x^{r} \, dx$, where $r < 0$. For which values of r will such integrals converge? Try some examples to check your answer.

5. Next try $\int_{1}^{2} \frac{dx}{\sqrt{x-1}}$. Can you obtain a closed form for $\int_{a}^{2} \frac{dx}{\sqrt{x-1}}$?

The next few exercises are exploratory, where the improper integral cannot be found by computing an integral in closed form.

6. Does the integral $\int_{0}^{1} \frac{1}{\sin(\sin(x))} \, dx$ converge or diverge? If it does converge, what is the limit?

7. Does the integral $\int_{0}^{1} \frac{1}{\sqrt{\sin(x)}} \, dx$ converge or diverge? If it does converge, what is the limit?

8. Does the integral $\int_{1}^{\infty} \frac{1}{\sin(\sin(x))} \, dx$ converge or diverge? If it does converge, what is the limit?

9. Does the integral $\int_{1}^{\infty} \frac{1}{\sqrt{\sin(x)}} \, dx$ converge or diverge? If it does converge, what is the limit?

4.7   Differential Equations

Prerequisites

- Some knowledge of differential equations

- The section on Bisection

Discussion

In this section we will apply our computer algebra system to the problem of solving first order ordinary differential equations. That is, we are seeking a function $y = y(x)$ that satisfies the equation

$$\frac{dy}{dx} = f(x, y)$$

Some examples are:

$$\frac{dy}{dx} = x^2, \quad \frac{dy}{dx} = \sin^3(x)(1+y)^3, \quad x\frac{dy}{dx} + (2+x)y = 2$$

where the third equation can be put in the form of the first equation by setting

$$f(x,y) = \frac{-(2+x)}{x}y + \frac{2}{x}$$

We can give a geometric interpretation to the problem of solving the differential equation $\frac{dy}{dx} = f(x,y)$ by recalling that $\frac{dy}{dx}$ is the slope of the curve $y = y(x)$ and thus $f(x,y)$ is the slope of the curve $y = y(x)$ at (x,y). If we draw a short segment of slope $f(x,y)$ at (x,y) for the first equation above, $\frac{dy}{dx} = x^2$, we obtain the following *direction field* (or *slope field*) for the equation:

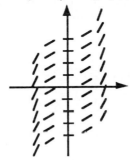

The solution $y = y(x)$ is a curve that, beginning from a starting value (x_0, y_0), has the desired slope at each point. Clearly, if f is a continuous function such a curve exists. For our example, if the starting point is $(x_0, y_0) = (0,0)$, so that $y(0) = 0$, then we are seeking the equation of the following curve:

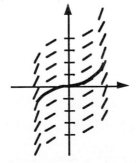

In some cases, the problem of solving a differential equation is exactly the same as performing an integration.

Example 1

We can find the solution to $\dfrac{dy}{dx} = x^4 \sin(x)$ by integration since the right-hand side depends only upon x. Integrating both sides,

$$\int \frac{dy}{dx}\, dx = \int x^4 \sin(x)\, dx$$

we have

$$y(x) = \int x^4 \sin(x)\, dx + C$$

where C is an arbitrary constant of integration. Applying our CAS

```
> y := int(x^4 *sin(x),x) + C;
```

we obtain

$$y(x) = -x^4 \cos(x) + 4x^3 \sin(x) + 12x^2 \cos(x) - 24 \cos(x) - 24x \sin(x) + C$$

as the general solution.

To obtain a specific solution, without the arbitrary constant, we need a condition. For example, if we are given that $y(0) = 1$, then we can solve for C

```
> evalf(solve(subs(x=0,y)=1,C));
```

obtaining $C = 25$. Thus the function

$$y(x) = -x^4 \cos(x) + 4x^3 \sin(x) + 12x^2 \cos(x) - 24 \cos(x) - 24x \sin(x) + 25$$

is the solution to the differential equation $\dfrac{dy}{dx} = x^4 \sin(x)$ with initial condition $y(0) = 1$. △

If $f(x, y)$ depends on both x and y, then one approach is *separation of variables*, where all the variables, including dx and dy are separated, those involving x on one side and those involving y on the other side of the equation so that we have $p(y)\, dy = q(x)\, dx$. We can then integrate both sides and try solving for y in terms of x. Note that if f depends on x or y alone, then separation of variables turns the problem into one of integration.

Example 2

We will solve the second example above, $\dfrac{dy}{dx} = \sin^3(x)(1 + y)^3$, by separation of variables. Dividing by $(1 + y)^3$ and multiplying by dx gives

$$\frac{dy}{(1 + y)^3} = \sin^3(x)\, dx$$

Now we apply integration to both sides

```
> int(1/(1+y)^3,y) = int(sin(x)^3,x) + C;
```

obtaining

$$-\frac{1}{2}\frac{1}{(1+y)^2} = -\frac{1}{3}\sin^2(x)\cos(x) - \frac{2}{3}\cos(x) + C$$

We can then solve for y,

```
> solve(",y);
```

to obtain the general solution. $\triangle$

If the differential equation $\dfrac{dy}{dx} = f(x,y)$ has the form

$$\frac{dy}{dx} + p(x)y = q(x)$$

then the equation is linear (in y) and we can find a solution by integration. We multiply by an *integrating factor*, $\exp(\int p(x)dx)$, obtaining

$$\frac{d}{dx}(y\exp(\int p(x)\,dx)) = q(x)\exp(\int p(x)\,dx)$$

We can now integrate both sides with respect to x and solve for y.

Example 3

Consider the differential equation

$$x\frac{dy}{dx} + (2+x)y = 2$$

Dividing both sides by x we have

$$\frac{dy}{dx} + \frac{(2+x)}{x}y = \frac{2}{x}$$

and obtain an integrating factor

```
> intfac:= exp(int((2+x)/x,x));
```

We can then solve for y:

```
> solve(intfac*y = int(intfac*2/x,x),y) + C;
```

Again, this gives us the general solution. We can find a particular solution as above, given some initial condition $y(x_0) = y_0$. △

If the differential equation can be written in the form

$$\frac{\partial F(x,y)}{\partial x} dx + \frac{\partial F(x,y)}{\partial y} dy = 0$$

for some function F, then the equation is called *exact*. In this case, the differential of F satisfies

$$dF(x,y) = \frac{\partial F(x,y)}{\partial x} dx + \frac{\partial F(x,y)}{\partial y} dy = 0$$

Thus $F(x,y) = C$ (an arbitrary constant) and the general solution of the differential equation can be found by solving $F(x,y) = C$ for y.

If we write the differential equation in the form

$$p(x,y) dx + q(x,y) dy = 0$$

we can test to see if the equation is exact. If $p = \dfrac{\partial F}{\partial x}$ and $q = \dfrac{\partial F}{\partial y}$, then since

$$\frac{\partial^2 F}{\partial y \partial x} = \frac{\partial^2 F}{\partial x \partial y}$$

for functions F with continuous second derivatives, we have

$$\frac{\partial p}{\partial y} = \frac{\partial^2 F}{\partial y \partial x} = \frac{\partial^2 F}{\partial x \partial y} = \frac{\partial q}{\partial x}$$

Thus a necessary condition for $p\, dx + q\, dy = 0$ to be exact is

$$\frac{\partial p}{\partial y} = \frac{\partial q}{\partial x}$$

Moreover, if this condition holds we can obtain F by integrating $\dfrac{\partial F}{\partial x} = p$ with respect to x:

$$F(x,y) = \int \frac{\partial F}{\partial x} dx + k(y) = \int p(x,y)\, dx + k(y)$$

for some function $k = k(y)$ depending on y alone. To determine $k(y)$, we take the derivative with respect to y:

$$\frac{dk}{dy} = -\frac{\partial}{\partial y} \int p(x,y)\, dx + \frac{\partial F}{\partial y} = -\frac{\partial}{\partial y} \int p(x,y)\, dx + q(x,y).$$

Finally, we integrate the above equation with respect to y, obtaining $k(y)$.

Example 4

Consider the differential equation

$$\frac{dy}{dx} = \frac{2x\sin(xy) + x^2\cos(xy)y}{-x^3\cos(xy) - 2y}$$

We can rewrite this as $p\,dx + q\,dy = 0$ where $p(x,y) = 2x\sin(xy) + x^2 y\cos(xy)$ and $q(x,y) = x^3\cos(xy) + 2y$. First we check that the equation is exact:

```
> p:= 2*x*sin(x*y) + x^2 *y*cos(x*y);
> q:= x^3 *cos(x*y) + 2*y;
> diff(p,y) - diff(q,x);
```

returns zero. Now we follow the above sequence of steps, simplifying intermediate results:

```
> intp:= simplify(int(p,x));
> ky:= int(Q - diff(intp,y),y);
> F:= intp + ky;
```

This returns the expression

$$F = x^2\sin(xy) + y^2$$

We cannot easily solve $x^2\sin(xy) + y^2 = C$ for y, but if we have an initial condition, for example $y(0) = 1$, then we can determine C by substituting $x = 0$, $y = 1$ into F,

```
> C:= subs({x=0,y=1},F);
```

This returns $C = 1$. We can also approximate the value of y at any point by substituting for x and then using approximation methods, e.g., the Bisection Algorithm discussed earlier in this text. For example, the value of y at $x = 2$ is approximated by substituting $x = 2$ in the equation $x^2\sin(xy) + y^2 = 1$ and defining the difference to be a function f of y:

```
> f:= proc(y) 4*sin(2*y) + y^2 - 1 end;
> bisect(f,0,1,0.001)
```

which returns a value of 0.121.   △

The equation may not be in the form required for the above methods, and even when the differential equation is in the correct form, the integration may be impossible to carry out (by woman or man or machine) in symbolic form. In this case, we cannot find the general solution but may be able to approximate the value of a particular solution at a point.

Suppose that we wish to approximate the value of the solution to

$$\frac{dy}{dx} = f(x, y)$$

at $x = x'$, given an initial condition $y(x_0) = y_0$. If x' is close to x_0, then we can approximate the value $y' = y(x')$ by

$$\frac{y' - y_0}{x' - x_0} = \frac{\Delta y}{\Delta x} \approx \frac{dy}{dx} = f(x_0, y_0)$$

Thus $y' \approx y_0 + f(x_0, y_0)(x' - x_0)$. Here we approximate by going along the tangent line to the curve at (x_0, y_0), which has equation

$$y = y_0 + f(x_0, y_0)(x - x_0)$$

rather than along the curve itself, $y = y(x)$.

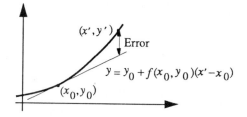

If x' is not close to x_0, then we can reach x' in several steps, going from x_0 to x_1, ..., $x_n = x'$. Thus for $k = 0$, ..., $n - 1$, the approximation y_n to $y(x_n)$ is obtained from computing the sequence of approximations

$$y_{k+1} = y_k + f(x_k, y_k)(x_{k+1} - x_k)$$

If we choose equally spaced values of x, $x_{k+1} - x_k = h$, then this method is called *Euler's Method*. We can use the CAS's **for** loop to do the computation.

Example 5

Suppose we wish to solve the differential equation

$$\frac{dy}{dx} = y^2 - x^4 - 2x^3 - x^2 + 2x + 1$$

with the initial condition that $y(0) = 0$. Suppose we want $y(1)$, the value of the solution y at $x = 1$. How small should h be? We know from our study of differential equations that the error is proportional to h (ignoring round-off errors). Thus to get one or two digits of accuracy, we will let $h = 0.01$. Now we use our CAS to define the function f, set the initial conditions, and compute the sequence of approximations using a **for** loop:

```
> f:= proc (x, y) y∧2 - x∧4 - 2*x∧3 - x∧2 + 2*x + 1 end;
> x:= 0;
> y:= 0;
> n:= 100;
> h:= 1/100;
> for k from 0 to n do
≫          print('x = ',x,'y = ',y);
≫          y:= evalf(y + f(x,y)*h);
≫          x:= evalf(x + h);
≫ od;
```

The final result printed is

$$x = 1.000000000, \ y = 1.966002277$$

How accurate is this? The exact solution to this equation is $y(x) = x^2 + x$, so $y(1) = 2$. Of course, usually we will not know the exact solution. In this case, we can estimate our error by recomputing with a smaller value of h and seeing if there is much change. For example, with $h = 0.02$, we obtain an approximation of $y = 1.982620981$. Thus we can be reasonably confident in the approximation to within about 0.1. Euler's Method is generally not very accurate, and round-off errors make it difficult to obtain high accuracy even when h is small. A more accurate approximation method is the Runge-Kutta Method, which has an error that is proportional to h^4. This is much more accurate than Euler's Method, since when h is small h^4 is much smaller than h.   △

Exercises

In the following problems, try to find the solution y to the differential equation with the given initial condition. If you cannot find the solution, then approximate the value of $y(1)$ to within 0.01.

1. $\dfrac{dy}{dx} = x\sin^3(x)$ with initial condition $y(0) = 0$

2. $\dfrac{dy}{dx} = \dfrac{3y}{x} + x\log(x)$ with initial condition $y(2) = 2$

3. $\dfrac{dy}{dx} = x^3 y + \cos(x)$ with initial condition $y(0) = 0$

4. $\dfrac{dy}{dx} = 3y + x^3$ with initial condition $y(0) = 1$

5. $\dfrac{dy}{dx} = x^2 \cos(x)y^2$ with initial condition $y(0) = 1$

6. $\dfrac{dy}{dx} = \sin(x^2)y^3$ with initial condition $y(0) = 1$

7. $\dfrac{dy}{dx} = \dfrac{\csc(y)\csc^2(x) - \exp(y^2)}{2xy\exp(y^2) - \csc(y)\cot(y)\cot(x)}$ with initial condition $y(\frac{\pi}{2}) = \dfrac{\pi}{4}$

8. $\dfrac{dy}{dx} = \dfrac{(x+1)(y+1)}{xy}$ with initial condition $y(2) = 2$

9. $\dfrac{dy}{dx} = -\dfrac{2x\sin(y) + y}{x^2\cos(y) + x + 8y}$ with initial condition $y(0) = 1$

10. $\dfrac{dy}{dx} = \sin(x^2 y^2)$ with initial condition $y(0) = 0$

Chapter 5

Series

5.1 Approximating Sums of Convergent Series

Prerequisites

- Convergence tests for series

- Section on Convergence of Sequences

Discussion

There are several convergence tests for series, the most common of which are the comparison, integral, ratio, and root tests. These tests can be used to determine if a series converges, but they do not tell us the limit. This section will examine the problem of approximating the value of an infinite series, once we know that it converges.

Integral Test Methods

If convergence is determined by the Integral Test, then the test also provides a means of estimating the error between a partial sum and the series value. Recall

Theorem 5.1.1 (The Integral Test) *If f is continuous and nonincreasing on the interval $[m, \infty)$, and if $a_k = f(k) \geq 0$ and is a nonincreasing sequence for all $k = m, m+1, m+2, \ldots$, then $\sum_{k=m}^{\infty} a_k$ converges if and only if $\int_m^{\infty} f(x)\, dx$ converges.*

Example 1

We can see that the series $\displaystyle\sum_{k=1}^{\infty} \frac{1}{k^3}$ converges by the integral test, since

$$\lim_{k\to\infty} \int_1^k \frac{1}{x^3}\, dx$$

exists. How can we approximate the value of $s = \displaystyle\sum_{k=1}^{\infty} \frac{1}{k^3}$ to within a given accuracy, say 0.001? The sequence of partial sums,

$$s_n = \sum_{k=1}^{n} \frac{1}{k^3}$$

will converge to the value s, so in order to determine how many terms to sum, i.e., the value of n to use, we need to estimate the error between s and s_n. This can be done from the integral test. The error between s and s_n is

$$|s - s_n| = \sum_{k=n+1}^{\infty} \frac{1}{k^3}$$

By examining the following diagram,

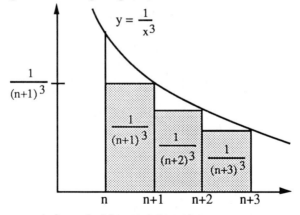

we see that the error is bounded by an integral:

$$|s - s_n| = \sum_{k=n+1}^{\infty} \frac{1}{k^3} \leq \int_n^{\infty} \frac{1}{x^3}\, dx = \lim_{b\to\infty}\left(\frac{1}{2n^2} - \frac{1}{2b^2}\right) = \frac{1}{2n^2}$$

If we want the error to be less than 0.001, then we need to find n satisfying

$$\frac{1}{2n^2} < 10^{-3}$$

or, $n \geq 23$. Thus the answer is

$$s_{23} = \sum_{k=1}^{23} \frac{1}{k^3}$$

Using our CAS we find that

> `sum(1/k∧3, k=1..23);`

returns the value of 1.201. △

Example 2

Sometimes the Integral Test does not apply directly, but can still be used. Since convergence of a series depends only on the behavior of the "tail" (the terms at the end), we can apply the Integral Test method to the tail. Consider the series $\sum_{k=1}^{\infty} \frac{1}{(k-10.5)^2}$. The terms $a_k = \frac{1}{(k-10.5)^2}$ will increase and then decrease, since the terms are reciprocals of a parabola $y = (x-10.5)^2$. We can see this using our CAS to print some of the terms:

```
> for k from 1 to 50 do
≫        print(k, 1/(k-10.5)∧2);
≫ od;
```

We see that the terms a_k decrease monotonically for $k \geq 12$. We can apply the Integral Test method to the tail:

$$\sum_{k=n+1}^{\infty} \frac{1}{(k-10.5)^2} \leq \int_{n}^{\infty} \frac{1}{(x-10.5)^2} \, dx = \frac{1}{n-10.5}$$

If we want to determine the sum to within 0.01, then we need $\frac{1}{(n-10.5)} \leq 0.01$ or $n \geq 111$. (Notice how the lower limit of integration reduces the rate of convergence!) Using our CAS we find that

> `sum(1/(k-10.5)∧2, k=1..111);`

returns the value of 9.76 to within 0.01. △

Example 3

Sometimes the integral in the Integral Test is not possible to evaluate exactly. In this case, estimates of the integral may be sufficient.

Suppose we wish to determine $\displaystyle\sum_{k=1}^{\infty} \frac{1}{(3k + \sin(k))^3}$ to within 10^{-4}. This series

can easily be seen to converge by comparing it to $\dfrac{1}{k^3}$:

$$\frac{1}{(3k + \sin(k))^3} \leq \frac{1}{(3k - 1)^3}$$

We know $\displaystyle\sum_{k=1}^{\infty} \frac{1}{(3x - 1)^3}$ converges. (Why?) We can use this comparison and the

integral comparison (as in the preceding examples) to estimate the tail of the series:

$$\sum_{k=n+1}^{\infty} \frac{1}{(3k + \sin(k))^3} \leq \sum_{k=n+1}^{\infty} \frac{1}{(3k - 1)^3} \leq \int_{n}^{\infty} \frac{1}{(3x - 1)^3} dx = \frac{1}{6(3n - 1)^2}$$

The error will be less than 10^{-4} if $6(3n - 1)^2 \geq 10^4$. Applying our CAS,

```
> solve(6*(3*n-1)^2 >= 10^4,n);
```

we find that $n \geq 14$. Thus s_{14} will give the desired accuracy and we use our CAS to compute

```
> evalf(sum(1/(3*n + sin(n))^3,n=1..14));
```

as 0.0236, to within 10^{-4}.   △

Example 4

WARNING: looking at the numerical values of the partial sums of a very slowly diverging series can mislead a student into erroneously believing the series converges. Consider the harmonic series, $\displaystyle\sum_{k=1}^{\infty} \frac{1}{k}$. This series is easily shown to be divergent by the integral test. However, it diverges very, very slowly. When looked at numerically, i.e., computing several partial sums, a student might easily believe that the series converges. For example, $s_{1500} - s_{700} = \displaystyle\sum_{k=1}^{1500} \frac{1}{k} - \sum_{k=1}^{700} \frac{1}{k} =$

0.761759232 and $s_{11,000} - s_{10,000} = \displaystyle\sum_{k=10,001}^{11,000} \frac{1}{k} = 0.015305635$. That is, a thousand terms add up to about 0.015. Such a slow rate of growth might very well lead an unquestioning student to conjecture that the series converges. ALWAYS determine whether or not a series converges *before* attempting to approximate its sum.   △

Alternating Series Methods

If the series is an alternating series, then we can also estimate the error. Recall the test for convergence of alternating series:

Theorem 5.1.2 (Alternating Series) *If the terms of the series* $\displaystyle\sum_{k=0}^{\infty}(-1)^k a_k$

satisfy the conditions

    (a) *All a_k's have the same sign (all positive or all negative)*

    (b) $|a_{k+1}| \leq |a_k|$ *for all k (terms are nonincreasing in absolute value)*

    (c) $\displaystyle\lim_{k\to\infty} a_k = 0.$

then the series $\displaystyle\sum_{k=0}^{\infty}(-1)^k a_k$ *converges and the difference between the limit and the nth partial sum is less than $|a_{n+1}|$. That is, if the limit is s, then*

$$\left| s - \sum_{k=0}^{n}(-1)^k a_k \right| \leq |a_{n+1}|$$

Example 5

We see that the series $\displaystyle\sum_{k=1}^{\infty}\frac{(-1)^k}{k^2}$ converges. If we want to determine the sum s to within 0.001, then we need to determine an n such that $|s - s_n| \leq 0.001$. From the above theorem, this will be true if the $n + 1$st term satisfies $|a_{n+1}| \leq 0.001$, i.e., $\dfrac{1}{(n+1)^2} \leq 0.001$, or $n \geq 31$. Thus $s_n = \displaystyle\sum_{k=1}^{31}\frac{(-1)^k}{k^2}$ will be within 0.001 of the sum s. Using our CAS we find that

```
> sum((-1)^k/k^2, k=1..31);
```

returns the value of -0.823. △

Heuristic Methods

A point plot of the sequence of partial sums of a convergent series can often be used to conjecture the sum of the series. The (conjectured) limit can then be "tested" by drawing a multipoint plot of the sequence of partial sums and the conjectured limit.

Example 6

We know that the series $\displaystyle\sum_{k=1}^{\infty}\frac{k!}{k^k}$ converges (apply the ratio test), but we do not know the value of its sum. We can observe from a point plot of the sequence of partial sums, $s_n = \sum\limits_{k=1}^{n} \dfrac{k!}{k^k}$, over [1,25] that the sum value is approximately 1.9. The multiplot below "tests" this conjecture. We use the Maple commands (the first command defines s as a function of n):

```
> s:= proc(n) sum(k!/(k∧k),k=1..n) end;
> plot({s, 1.9},1..25, style=POINT);
```

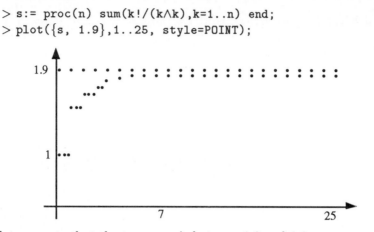

This plot suggests that the true sum is between 1.8 and 1.9.   △

Another heuristic method is to use a CAS to compute several partial sums. When the partial sums stop changing, within our desired accuracy, then we assume we are sufficiently close to the answer. This assumption may not be justified, but it may be the best that we can do.

Example 7

Consider the series $\sum\limits_{k=1}^{\infty} \dfrac{1}{k!}$. It clearly converges (from the ratio test), but what is its sum to within 10^{-5}? We cannot use the integral estimate, but the terms $a_k = \dfrac{1}{k!}$ decrease rapidly and monotonically. Thus we expect good behavior when we compute the partial sums, and stop when they stop changing. We use our CAS, and print out a range using a **for** loop.

```
> for n from 1 to 20 do
≫         print(n, evalf(sum(1/(k!), k=1..n)));
≫ od;
```

a portion of the output is

$$9, 1.718281526$$
$$10, 1.718281801$$
$$11, 1.718281826$$
$$12, 1.718281828$$

Thus, with an accuracy of 10^{-6}, the sum appears to be 1.718282. (Do you recognize this number?)   △

Exercises

In each of the following,

> **(a)** Determine if the series converges.
>
> **(b)** Explain why it does (or does not) converge.
>
> **(c)** If it does converge, find the limit to within 0.001.

1. $\sum_{k=1}^{\infty} \frac{1}{k^2}$

2. $\sum_{k=1}^{\infty} \frac{1}{k^5}$

3. $\sum_{k=1}^{\infty} \frac{1}{k}$

4. $\sum_{k=1}^{\infty} \frac{1}{k^4}$

5. $\sum_{k=1}^{\infty} \frac{1}{(k - 10/2)^2}$

6. $\sum_{k=1}^{\infty} \frac{(-1)^{k-1}}{k^5}$

7. $\sum_{k=1}^{\infty} \frac{1}{(2k + \cos(k) + 1)^4}$

8. $\sum_{k=1}^{\infty} \frac{1}{(k^3 + 2k + 8)^4}$

9. $\sum_{k=1}^{\infty} \frac{(-1)^k}{(2k + \cos(k) + 1)^4}$

10. $\sum_{k=1}^{\infty} \frac{(-1)^k}{(k - 8/3)^3}$

11. $\sum_{k=1}^{\infty} \frac{(-1)^k}{k^4}$

12. $\sum_{k=1}^{\infty} \frac{(-1)^{k+1}1}{k^3}$

13. $\sum_{k=1}^{\infty} \frac{k^2}{k!}$

14. $\sum_{k=1}^{\infty} \frac{1}{2^{k^2}}$

15. $\sum_{k=1}^{\infty} \frac{2^k}{k!}$

16. $\sum_{k=1}^{\infty} \frac{\sin((-1)^k)}{\sin(k) + k^2}$

17. $\sum_{k=3}^{\infty} \frac{1}{k \ln(k)}$

18. $\sum_{k=3}^{\infty} \frac{1}{k^2 \ln(k)}$

19. Consider the series $\sum_{k=1}^{\infty} \frac{1}{k \sin(k^2)}$. The partial sums for this series for $n = 35$ to 80 in steps of 5 are given in the following table.

n	s_n	n	s_n
35	-4.028280877	60	-4.029502567
40	-4.135168053	65	-4.036982636
45	-4.179512035	70	-4.045021033
50	-4.274527714	75	-4.027283415
55	-4.201255317	80	-4.028280877

Do these partial sums indicate that the series converges to a number close to -4? Why? Determine if the series converges. If it does, can you approximate its sum with an accuracy of 10^{-3}? Why? If the series diverges, are the partial sums unbounded? Why?

20. The purpose of this exercise is to develop a "feeling" for the behavior of the *harmonic series,* $\sum_{k=1}^{\infty} \frac{1}{k}$.

(a) Is the sequence of partial sums, s_n, an increasing sequence? Why?

(b) Compute s_n for $n = 10, 10^{10}, 10^{100}, 10^{1,000}, 10^{10,000}$. Conjecture how s_n changes when n increases by a power of 10.

(c) How does the rate of growth of s_n compare with the rate of growth of $\log(n)$?

(d) Show that $\log(n) + \frac{1}{n} < s_n < 1 + \log(n)$. Hint: Think of $\log(n)$ as the area under the graph of $f(x) = \frac{1}{x}$ over the interval $[1, n]$. Draw pictures showing the upper and lower Riemann Sums.

(e) Using $\log(n)$ to approximate s_n, determine a value of n such that $s_n > 100$. Does using $\log(n)$ to approximate s_n help explain the pattern that you observed in part b? Explain.

21. The purpose of this exercise is to develop a "feeling" for the behavior of the *alternating harmonic series,* $\sum_{k=1}^{\infty} (-1)^{k+1} \frac{1}{k}$.

(a) Compute the sequence of partial sums, s_n, for $n = 2, 4, 6, 8, 10$. Is $\{s_{2n}\}$ an increasing sequence? Prove your assertion.

(b) Compute s_n for $n = 1, 3, 5, 7, 9$. Is $\{s_{2n+1}\}$ a decreasing sequence? Prove your assertion.

(c) What relationship (if any) exists between $\{s_{2n}\}$ and $\{s_{2n+1}\}$? Explain.

(d) Assume $\sum_{k=1}^{\infty} (-1)^{k+1} \frac{1}{k}$ converges and has sum L. What relationship exists between $\{s_{2n}\}, \{s_{2n+1}\}$, and L? Use this relationship to obtain an error bound for $|\sum_{k=1}^{\infty} (-1)^{k+1} \frac{1}{k} - s_{20}|$. Explain your reasoning.

5.2   Taylor's Theorem

Prerequisites

- Taylor's Theorem for functions of one variable

- The section on Elementary Graphing

- An example and some exercises use the section on Improper Integrals

Discussion: Part I

Taylor's Theorem says that if $f : \Re \longrightarrow \Re$ has $n+1$ continuous derivatives on an interval containing c and x, then f is equal to a (Taylor) polynomial "expanded about c" or "in powers of $x - c$" plus an error term,

$$f(x) = \sum_{k=0}^{n} \frac{f^{(k)}(c)}{k!} (x - c)^k + R_n(x)$$

where the error term

$$R_n(x) = \frac{f^{(n+1)}(d)}{(n+1)!} (x - c)^{n+1}$$

for some d between c and x.

This theorem gives us an error bound for approximating "smooth" functions (those having $n+1$ continuous derivatives) by polynomials.

Most CASs have a Taylor polynomial command. The Maple CAS *taylor* command also gives an order of magnitude term for the error term using the Big 0 notation. For example, to obtain the nth degree Taylor polynomial for $\frac{1}{1-x}$ expanded about $x = 0$ with an order of magnitude term, one enters

```
> taylor(1/(1 - x), x = 0,n+1);
```

The CAS Maple responds with

$$1 + x + x^2 + x^3 + \cdots + x^n + O(x^{n+1})$$

(The Big 0 notation in the last term indicates the order of magnitude of the remainder. In this example, "$O(x^{n+1})$" means that there exists a constant $C > 0$ such that for large x, $s_n < C x^{n+1}$. See the section on Growth of Functions.) To extract the nth degree polynomial, s_n, we use the **convert** command with the parameter **polynom**.

```
> convert(taylor(1/(1 - x), x = 0, n+1),polynom);
```

$$1 + x + x^2 + x^3 + \cdots + x^n$$

Note that $n + 1$ is entered into the command to obtain a polynomial of degree n.

Example 1

Suppose we wish to approximate $\sin(1)$ to within 0.001. The sine function has derivatives of all orders on all of $\Re$, so the conditions of the theorem are satisfied.

First we choose a point c at which to evaluate the function and its derivatives. Since the trigonometric functions are most easily evaluated at zero, we will choose $c = 0$. Since $x = 1$ we have

$$\sin(1) = \sum_{k=0}^{n} \frac{\sin^{(k)}(0)}{k!} (1)^k + R_n(1)$$

where $R_n(1) = \dfrac{\sin^{(n+1)}(d)}{(n+1)!} (1)^{n+1}$ for some d between 0 and 1.

Next we need to determine the number of terms needed to obtain the desired accuracy. We need to find an n so that

$$|R_n(1)| \leq 0.001$$

that is,

$$\left| \frac{\sin^{(n+1)}(d)}{(n+1)!} (1)^{n+1} \right| \leq 0.001$$

Since $|\sin^{(n+1)}(d)| \leq 1$ for any n and d, we need

$$\frac{1}{(n+1)!} \leq 0.001 \quad \text{or} \quad (n+1)! \geq 1000$$

Evaluating some factorials (with the aid of our CAS) we find:

$$\begin{aligned} 6! &= 720 \\ 7! &= 5040 \end{aligned}$$

Thus $n + 1 = 7$ or $n = 6$ and so by Taylor's Theorem,

$$\left| \sin(1) - \sum_{k=0}^{6} \frac{\sin^{(k)}(0)}{k!} \right| \leq 0.001$$

We evaluate $\displaystyle\sum_{k=0}^{6} \frac{\sin^{(k)}(0)}{k!}$ using our CAS. There are 3 steps involved: first, obtain a sixth degree Taylor polynomial for $\sin(x)$ expanded about $x = 0$; second, substitute $x = 1$ into the polynomial; third, evaluate the expression to decimal form. We combine these three steps into one command.

```
> evalf(subs(x=1,convert(taylor(sin(x),x=0,7),polynom)));
```

$$0.8416666667$$

An approximation accurate to 10 decimal places is 0.8414709848.   △

The next step is to approximate a function f over an interval containing a specified point, rather than just at the point. This will involve using both Taylor's polynomial and the error term, and is done in the section on Approximation by Taylor's Polynomials.

Exercises

1. Find an approximation to $\cos(2)$ with an error of at most 0.001.

2. Find an approximation to e^2 with an error of at most 0.001.

Discussion: Part II

In the preceding section, we used Taylor's Theorem to approximate a function value at a point. In this section, we will use Taylor's Theorem to determine the behavior of a function near a given point. We wish to compare the behavior of a (complicated) function to a polynomial, such as $g(x) = a + b(x - c)^n$ where c is the given point, and a and b are parameters. We know what these functions look like, so they are a good standard of comparison. (See the section on Elementary Graphing for a discussion of these functions.) We use Taylor's Theorem to find the desired polynomial.

Example 2

In the section on Limits of Functions, we were interested in the behavior of $f(x) = \dfrac{x^3 \sin(x) - 7x}{x \cos(x)} + 7$ near $x = 0$. We now wish to find a function $g(x) = a + b(x - 0)^n = a + bx^n$ that behaves like f near 0. This is a job for Taylor's theorem.

There are several possible approaches. Having studied Taylor's Theorem, we know the Taylor expansion for standard functions like sin and cos near 0: $\sin(x) = x - x^3/3! + O(x^4)$ and $\cos(x) = 1 - x^2/2! + O(x^3)$, using the "Big O" terms which denote the order of magnitude of the error terms. We can simplify f and then substitute the first two terms in place of sin and cos:

$$
\begin{aligned}
f(x) &= \frac{x^3 \sin(x) - 7x}{x \cos(x)} + 7 = \frac{x^2 \sin(x) - 7 + 7\cos(x)}{\cos(x)} \\
&\approx \frac{x^2(x - x^3/6) - 7 + 7(1 - x^2/2)}{1 - x^2/2} = \frac{x^3 - x^4/6 - 7x^2/2}{1 - x^2/2}
\end{aligned}
$$

Now, when x is near zero high powers of x are very small. Thus in the above expression we eliminate the higher powers of x from both the numerator and denominator, obtaining $f(x) \approx -7x^2/2$. Thus f behaves like $g(x) = -7x^2/2$ near $x = 0$.

We illustrate the approximation by graphing both the function, f, and its approximation, g, near 0:

> plot({(x^3*sin(x)-7*x)/(x*cos(x))+7, -7*x^2/2},x=-1..1);

returns

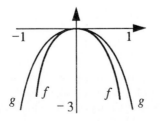

If the function f were more difficult to analyze, we might want to use the CAS to compute the first few terms of the Taylor's polynomial for f near the point. Rather than constructing the sum as in the previous example, we use our CAS to obtain the fourth degree Taylor polynomial expanded about $x = 0$.

```
> convert(taylor((x^3 *sin(x)-7*x)/(x*cos(x))+7, x=0,5),polynom;
```

$$\frac{-7}{2}x^2 + x^3 - \frac{35}{24}x^4$$

$\triangle$

Example 3

In the section on Improper Integrals we considered the problem of convergence of integrals like $\int_0^1 \frac{1}{\sin(\sin(x))}\,dx$. How can we tell if it converges? We cannot compute a closed form expression for this integral, but we can approximate $\frac{1}{\sin(\sin(x))}$ near zero by a rational function whose behavior is understood. We apply the **taylor** command

```
> taylor(sin(sin(x)),x=0,5);
```

$$x - x^3/3 + O(x^5)$$

This says that $\sin(\sin(x))$ behaves like $x - x^3/3$ near 0, where terms of degree 5 or more are neglected. As a first degree approximation, $y = \sin(\sin(x))$ behaves like $y = x$ near 0; thus $\frac{1}{\sin(\sin(x))}$ behaves like $\frac{1}{x}$ near 0. Since $\int_0^1 \frac{1}{x}\,dx$ diverges, we expect $\int_0^1 \frac{1}{\sin(\sin(x))}\,dx$ to diverge also. As a check on our reasoning, we can use the **limit** command to compute $\lim\limits_{y \to 0} \int_y^1 \frac{1}{\sin(\sin(x))}\,dx$:

```
> limit(int(1/sin(sin(x)),x=y..1),y=0);
```

returns "infinity."

$\triangle$

In general, given an integral, $\int_c^b f(x)\,dx$, that is improper near $x = c$ and given that the integrand, $f(x)$, "behaves like" $a(x - c)^n$ near $x = c$, then the convergence of the integral can be tested by checking the convergence of the integral of $a(x - c)^n$. See exercises $8 - 10$.

Exercises

In Exercises 3 through 6 find a function $g(x) = a + b(x - c)^n$ that behaves like f near c. Use Taylor's Theorem to find g; then check by graphing.

3. $f(x) = \dfrac{x^2 \sin^2(x) \cos(x)}{(x + 1)^3}$, $c = 0$.

4. $f(x) = \dfrac{e^x \sin(x) \cos(x)}{x^2}$, $c = 1$.

5. $f(x) = (x \sin(x) + 2 \cos(x))^2$, $c = 0$.

6. $f(x) = \tan(x^2) \sin(x^2) + \cos(x)x^2$, $c = 1$.

7. What condition on f will guarantee that f will behave like $g(x) = a + b(x - c)^n$ for $n = 1$ (rather than some other power) near $x = c$?

8. Determine whether or not $\int_0^1 \cot(x)\,dx$ converges.

9. Determine whether or not $\int_0^1 \sqrt{\cot(x)}\,dx$ converges.

10. Determine whether or not $\int_{-1}^1 1/\sqrt{1 + x^3}\,dx$ converges.

5.3 Approximation by Taylor's Polynomials

Prerequisites

- Section on Taylor's Theorem

Discussion

In the section on Taylor's Theorem, we approximated a function value at a point using Taylor's Theorem. Now we use this theorem to approximate a function by a Taylor polynomial over a given interval. Recall that Taylor's Theorem says that if a function f has $n + 1$ continuous derivatives on an interval containing c and x, then

$$f(x) = \sum_{k=0}^{n} \frac{f^{(k)}(c)}{k!} (x - c)^k + R_n(x)$$

where

$$R_n(x) = \frac{f^{(n+1)}(d)}{(n + 1)!} (x - c)^{n+1}$$

for some d between c and x.

Our approximating polynomial will be

$$P_n(x) = \sum_{k=0}^{n} \frac{f^{(k)}(c)}{k!} (x - c)^k$$

What we must do is bound the error estimate, $R_n(x)$, not just at a single value, but *over an interval*.

A typical problem is to determine a polynomial that approximates a given function over the interval $[a, b]$ within a prescribed accuracy, denoted by ACC. The general solution procedure for such a problem is:

1. Pick a point c at which to evaluate f and its derivatives. If we do the computation by hand, then we need to pick a point c where the derivatives $f^{(n)}(c)$ are easy to compute. However, since we have a CAS at our disposal we can pick a point c to make the degree of the polynomial small. This will usually be the midpoint of the given interval, $c = \dfrac{a+b}{2}$.

2. Determine n so that

$$|R_n(x)| \le ACC \text{ for all } x \text{ in } [a, b]$$

3. Compute P_n for the value of n determined above. Then

$$|f(x) - P_n(x)| \le ACC \text{ for all } x \text{ in } [a, b]$$

Thus P_n will be the desired polynomial approximation to f on $[a, b]$.

Fxample

Ve will approximate $f(x) = \sin^2(x)$ on the interval $[2, 3]$ by two Taylor polynomials with an accuracy, $ACC = 0.001$. For one polynomial, we choose c as a point where $f^{(n)}(c)$ is easy to evaluate and for the other polynomial, we choose c as the midpoint of the interval, $c = 2.5$.

Part I

Where is $f^{(n)}(c)$ easy to evaluate? Since $f^{(n)}$ is a product of sines and cosines, it is easy to evaluate at multiples of $\pi/2$. The closest integral multiple of $\pi/2$ to our interval is π. Thus we choose $c = \pi$ and find an n such that

$$R_n(x) = \frac{f^{(n+1)}(d)}{(n+1)!}(x - c)^{n+1} \le 0.001$$

for all x in $[2, 3]$ and d in $[2, \pi]$. Well,

$$|R_n(x)| = |\frac{f^{(n+1)}(d)}{(n+1)!}(x - c)^{n+1}| = |\frac{f^{(n+1)}(d)}{(n+1)!}| |x - c|^{n+1}$$

The second factor is easy to handle, since $|x - c| = |x - \pi| \le |2 - \pi|$. The first factor requires us to estimate $|f^{(n+1)}(d)|$, where d is some unknown point in $[2, \pi]$. We use our CAS to compute some values of $f^{(k)}$, hoping to see a pattern:

```
> f:= sin(x)∧2;
> for k from 1 to 10 do
≫        print(k,diff(f,x$k)) # Diff.  f k times.
≫ od;
```

This returns

$$
\begin{array}{cr}
1, & 2\sin(x)\cos(x) \\
2, & 2\cos(x)^2 - 2\sin(x)^2 \\
3, & -8\sin(x)\cos(x) \\
4, & -8\cos(x)^2 + 8\sin(x)^2 \\
5, & 32\sin(x)\cos(x) \\
6, & 32\cos(x)^2 - 32\sin(x)^2 \\
7, & -128\sin(x)\cos(x) \\
8, & -128\cos(x)^2 + 128\sin(x)^2 \\
9, & 512\sin(x)\cos(x) \\
10, & 512\cos(x)^2 - 512\sin(x)^2
\end{array}
$$

We recognize the pattern (after a little experimenting)

$$
f^k(x) = \begin{cases} (-1)^{n+1}2^k \sin(x)\cos(x) & \text{for } k = 2n-1 \\ (-1)^{n+1}[2^{k-1}\cos^2(x) - \sin^2(x)] & \text{for } k = 2n \end{cases}
$$

(A challenging subproblem is to prove this result by Mathematical Induction.)

Since $|\sin(x)| \le 1$ and $|\cos(x)| \le 1$ for all x, and $|r - s| \le |r| + |s|$ for any real numbers, we see from the table that $|f^{(k)}(x)| \le 2^k$. Thus we have

$$
|R_n(x)| \le \left| \frac{f^{(n+1)}(d)}{(n+1)!} \right| |2 - \pi|^{n+1} \le \frac{2^{n+1}}{(n+1)!}(\pi - 2)^{n+1}
$$

Now we find an n so that

$$
\frac{2^{n+1}}{(n+1)!}(\pi - 2)^{n+1} \le 0.001
$$

We can use a **for** loop to print out some values. After some experimentation with ranges of n,

```
> for n from 7 to 11 do
≫        print(n, evalf(2∧(n+1)*(Pi-2)∧(n+1)/(n+1)!))
≫ od;
```

returns

$$
\begin{array}{cl}
7, & .01831507773 \\
8, & .004646301819 \\
9, & .001060836806 \\
10, & .0002201897281 \\
11, & .00004189449604
\end{array}
$$

Thus the degree of the polynomial must be $n = 10$.

Finally we use the **convert taylor** with parameter **polynom** command (see the section on Taylor's Theorem).

> `convert(taylor(f,x=Pi,11), polynom);`

$$(x - \text{Pi})^2 - 1/3 \, (x - \text{Pi})^4 + 2/45 \, (x - \text{Pi})^6 - 1/315 \, (x - \text{Pi})^8 + 2/14175 \, (x - \text{Pi})^{10}$$

This is our approximating polynomial when $c = \pi$.

Part II

Now we choose $c = 2.5$. Thus we want

$$|R_n(x)| \leq |\frac{f^{(n+1)}(d)}{(n+1)!}| \, |2 - 2.5|^{n+1} \leq \frac{2^{n+1}}{(n+1)!} (0.5)^{n+1} = \frac{1}{(n+1)!} \leq 0.001$$

Since this last inequality is satisfied for $n \geq 6$, our approximating polynomial is

$$.3581689073 - .9589242746(x - 2.5) + .2836621853(x - 2.5)^2 + .6392828498(x - 2.5)^3$$
$$- .09455406175(x - 2.5)^4 - .1278565699(x - 2.5)^5 + .01260720823(x - 2.5)^6$$

where the coefficients have been approximated to 10 digits.   △

Note that choosing c to be the midpoint of the interval led to an approximating polynomial of degree 6 compared to degree 10 when the value of c was chosen in order to simplify the computation of $f^{(n)}$.

Exercises

1. Find a polynomial with rational coefficients approximating $f(x) = \sin^2(x)$ to within 0.001 over the interval $[2, 3]$.

2. Find a polynomial with rational coefficients approximating $f(x) = \sin^2(x)$ to within 0.001 over the interval $[1, 2]$.

3. Find a polynomial with rational coefficients approximating $f(x) = e^x$ to within 0.01 over the interval $[2, 3]$.

4. Find a polynomial with rational coefficients approximating $f(x) = e^x$ to within 0.01 over the interval $[1, 2]$.

5. Find a polynomial approximating $f(x) = e^x$ to within 0.01 over the interval $[2, 3]$ if we wish to minimize the degree of the polynomial.

6. Find a polynomial approximating $f(x) = e^x$ to within 0.01 over the interval $[1, 2]$ if we wish to minimize the degree of the polynomial.

5.4   Convergence of Power Series

Prerequisites

- Convergence of sequences.

- The section on Taylor's Theorem

- Definition of power series, partial sum, and convergence of power series

Discussion

A *power series* is a series of the form

$$\sum_{k=0}^{\infty} a_k(x-c)^k = a_0 + a_1(x-c) + a_2(x-c)^2 + \cdots$$

Truncating a series after $n+1$ terms leaves an nth degree polynomial called the nth partial sum of the series and denoted by s_n,

$$s_n = \sum_{k=0}^{n} a_k(x-c)^k$$

A power series converges if and only if the corresponding sequence of partial sums, $\{s_n\}$, converges.

A Taylor's series for f "expanded about c" or "in terms of powers of $(x-c)$" is a power series in which the coefficients a_k are given by

$$a_k = \frac{f^{(k)}(c)}{k!}$$

The series converges on an interval, called the *interval of convergence* centered at c. (See the section on Radius of Convergence.) A Maclaurin series is a Taylor's series with $c = 0$.

Most CASs have a built-in Taylor series command for obtaining an nth partial sum. The Maple CAS *taylor* command also gives an order of magnitude term using the Big O notation. For example to obtain the nth partial sum, s_n, of the Taylor series for $\dfrac{1}{1-x}$ expanded about $x = 0$ with an order of magnitude term, one enters

```
> taylor(1/(1 - x), x = 0,n+1);
```

The CAS Maple responds with

$$1 + x + x^2 + x^3 + \cdots + x^n + O(x^{n+1})$$

(The Big O notation in the last term indicates the order of magnitude of the remainder. In this example, "$O(x^{n+1})$" means that there exists a constant $C > 0$ such that for large x, $s_n \le Cx^{n+1}$.) To extract the nth degree polynomial, s_n, we use the **convert** command with the parameter **polynom**.

```
> convert(taylor(1/(1 - x), x = 0, n+1),polynom);
```

$$1 + x + x^2 + x^3 + \cdots + x^n$$

Note that $n + 1$ is entered into the command to obtain a polynomial of degree n.

Since convergence of a power series is expressed in terms of the sequence of partial sums, questions concerning convergence, approximations, and error bounds are central to the analysis of power series just as they are for sequences of constants. The major difference in the analysis of a power series compared to a sequence of constants is that each term of the sequence of partial sums of a power series is a polynomial function, and thus we need to restrict the domains of these functions to the interval of convergence of the series.

In this section, we shall use a CAS to graphically illustrate convergence and the interval of convergence. We shall also compare worst case error bounds given by Taylor's Remainder Theorem and the Error Formula for Alternating Series against the exact errors obtained through graphical analysis.

Example 1

We use a CAS to superimpose the graphs of the first five partial sums of the geometric series for $\dfrac{1}{1 - x}$ expanded about $x = 0$ onto the graph of $y = \dfrac{1}{1 - x}$.
We begin by defining s_n as a function of n.

```
> s:= proc(n) convert(taylor(1/(1-x),x=0,n+1),polynom) end;
```

Using a **for** loop, we display the first five partial sums. (Remember that the value of $n + 1$ in the command must be one greater than the desired degree of the polynomial.)

```
> for n from 0 to 4 do
≫ print(s(n));
≫ od;
```

$$1$$
$$1 + x$$
$$1 + x + x^2$$
$$1 + x + x^2 + x^3$$
$$1 + x + x^2 + x^3 + x^4$$

We now enter the command for a multiplot of $s_0, s_1, s_2, s_3, s_4,$ and $\dfrac{1}{1 - x}$.

```
> plot({s(0),s(1),s(2),s(3),s(4),1/(1-x)},x=-2..2);
```

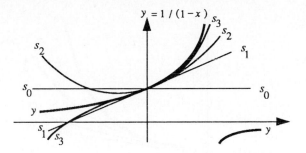

Note how the graphs of the partial sums, s_0 through s_4, appear to "converge" to the graph of $y = \dfrac{1}{1-x}$ for x near $x = 0$, but to diverge from it for $x < -1$ and $x > 1$. This confirms our understanding that the interval of convergence of the geometric series for $y = \dfrac{1}{1-x}$ is $-1 < x < 1$.   △

Example 2

We superimpose the graphs of the first four partial sums for the series expansion of e^x about $x = 0$ onto the graph of $f(x) = e^x$ over the interval $[-1, 2]$. As in Example 1, we define the nth partial sum as a function of n and then generate the first four partial sums.

```
> s:= proc(n) convert(taylor(exp(x),x=0,n+1),polynom) end;
> for n from 0 to 3 do
≫ print(s(n));
≫ od;
```

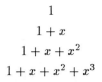

To obtain the multiplot, we enter

```
> plot({s(0),s(1),s(2),s(3),exp(x)},x=-1..2);
```

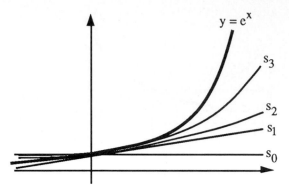

Note how the convergence of s_n to e^x "spreads outward" from the center, $x = 0$, as n increases. For example, considering only positive values of x and an error bound of 0.5, we have

$$
\begin{aligned}
|e^x - s_0| &= |e^x - 1| < 0.5 \text{ for } 0 < x < 0.41 \\
|e^x - s_1| &= |e^x - 1 - x| < 0.5 \text{ for } 0 < x < 0.86 \\
|e^x - s_2| &= |e^x - 1 - x - x^2/2| < 0.5 \text{ for } 0 < x < 1.28 \\
|e^x - s_3| &= |e^x - 1 - x - x^2/2 - x^3/6| < 0.5 \text{ for } 0 < x < 1.69
\end{aligned}
$$

We now compare the error bound given by Lagrange's Remainder Theorem to the maximum actual error. The expression for the error bound is

$$
R_n = |f(x) - s_n| \leq \frac{f^{(n+1)}(d)}{(n+1)!}(x - c)^{(n+1)}
$$

where d is some value between x and c. In this example, $f(x) = e^x$ and $c = 0$. Thus the error bound when $n = 3$ is $\frac{e^d}{24}x^4$. Since the problem specified the interval $[-1, 2]$ and e^x is an increasing function, the *worst case* error bound is $\frac{e^2}{24}2^4$ or 4.9260.

It is clear from the graphs that when approximating e^x by s_3 the maximum actual error is $|e^x - s_3|$ evaluated at $x = 2$. (Note that $\{s_n\}$ is an increasing sequence.) Thus the maximum actual error is 1.0557.   △

Example 3

We repeat the process illustrated in Examples 1 and 2 for the natural logarithm function, denoted by $\log(x)$, centered at $c = 1$ with the first five partial sums. Thus we define the nth partial sum as a function of n and generate the partial sums for $n = 0$ through 4.

```
> s:= proc(n) convert(taylor(log(x),x=1,n+1),polynom) end;
> for n from 0 to 4 do
```

```
>> print(s(n));
>> od;
```

$$0$$
$$x - 1$$
$$x - 1 - 1/2(x - 1)^2$$
$$x - 1 - 1/2(x - 1)^2 + 1/3(x - 1)^3$$
$$x - 1 - 1/2(x - 1)^2 + 1/3(x - 1)^3 - 1/4(x - 1)^4$$

Note that the terms in the partial sums alternate in sign in contrast to the situation in the two previous examples. (The Taylor's series for $g(x) = \log(x)$ is an alternating series.)

We omit s_0 (since $s_0(x) = 0$) and use our CAS to sketch the remaining four curves.

```
> plot({s(1),s(2),s(3),s(4),log(x)},x=0.4..3);
```

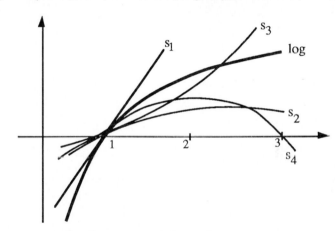

We now compare the three error statements:

1. Maximum actual error

2. Worst case form of Lagrange's Remainder Theorem

3. Alternating Series Error Formula

From the graphs, it is clear that when approximating $\log(x)$ by s_4 the maximum actual error is $|\log(x) - s(4)|$ evaluated at $x = 3$. Using our CAS as a calculator, we use the evaluation command

```
> evalf(log(s) - subs(x=3,s(4)));
```

to obtain the *actual* error of 2.43195622.

Lagrange's Remainder Theorem says that for some d between x and 1 (recall that $g(x) = \log(x)$)

$$R_4 = |g(3) - s_4(x)| \leq \frac{g^{(5)}(d)}{5}(x - 1)^5$$

We use the substitute command of our CAS to evaluate the fifth derivative of $g(x)$ at $x = d$.

```
> subs(x=d,diff(log(x),x$5));
```

$$\frac{24}{d^5}$$

Thus we have

$$R_n = |g(x) - s_4| \leq \frac{24}{120 d^5}(x - 1)^5$$

The worst case is obtained by setting $d = 0.4$ and $x = 3$ which yields an error bound of 625. This is *much* greater than the actual error.

The Alternating Series Error Formula says that

$$|g(x) - s_4(x)| \leq \frac{1}{5}(x - 1)^5 \leq \frac{32}{5} = 6.4$$

Quite an improvement over Lagrange's error estimate!   △

Exercises

In Exercises 1 through 8, find the indicated partial sums of the power series expanded about c of the given function. Superimpose the graphs of the indicated partial sums onto the graph of the function. Compare the error bounds given by Lagrange's Remainder Theorem and (where appropriate) the Error Formula for Alternating Series for the maximum actual error.

1. $f(x) = \tan(x)$ for $-2 \leq x \leq 2$; $c = 0$. All partial sums of degree ≤ 5

2. $f(x) = \cosh(x)$ for $-2 \leq x \leq 2$; $c = 0$. All partial sums of degree ≤ 4

3. $f(x) = (1 + x^2)^{-1/2}$ for $-2 \leq x \leq 2$; $c = 0$. All partial sums of degree ≤ 4

4. $f(x) = \sqrt{1 + x^2}$ for $0 \leq x \leq 2$; $c = 0$. All partial sums of degree ≤ 6

5. $f(x) = \cos(x)$ for $-4 \leq x \leq 4$; $c = 0$. Partial sums of degrees 6 $-$ 10

6. $f(x) = \cos(x)$ for $-7 \leq x \leq 7$; $c = 0$. Partial sums of degrees 14 and 16

7. $f(x) = \sin(x)$ for $-2 \leq x \leq 6$; $c = 2$. Partial sums of degrees 5 $-$ 8

8. $f(x) = \log(x)$ for $0.5 \leq x \leq 5$, $c = 2$. All partial sums of degree ≤ 5

9. If all the terms in a power series for the function f are positive, where will the graphs of the partial sums lie with respect to the graph of f (above, below, or neither)? Why?

10. Show that the alternating harmonic series converges to $\log(2)$. Hint: Expand $\log(x)$ in a Taylor series about $x = 1$. Determine n such that s_n for the alternating harmonic series approximates $\log(2)$ with accuracy less than or equal to 0.01.

5.5 Radius of Convergence

Prerequisites

- An introduction to the radius of convergence

- The section on Convergence of Power Series

Discussion

Consider a power series in x centered at c with coefficients $\{a_k\}$:

$$\sum_{k=0}^{\infty} a_k (x - c)^k$$

As you have learned, this series converges absolutely on an interval centered at c of radius R, i.e., $(c - R, c + R)$. R is called the *radius of convergence*. In the section on Convergence of Power Series, we illustrated finding the radius of convergence of a series by looking at a multigraph of the first few partial sums of the series. In this section we will investigate how to determine the radius of convergence analytically.

There are two convergence tests that are natural to apply: the root test and the ratio test.

Applying the root test, if

$$\lim_{k \to \infty} |a_k (x - c)^k|^{1/k} = \lim_{k \to \infty} |a_k|^{1/k} |(x - c)| < 1$$

then the series will converge absolutely for all x satisfying

$$|x - c| < 1/ \lim_{k \to \infty} |a_k|^{1/k}$$

Thus the radius of convergence is at least $1/ \lim_{k \to \infty} |a_k|^{1/k}$.

Applying the ratio test, if

$$\lim_{k \to \infty} \left| \frac{a_{k+1}(x - c)^{k+1}}{a_k(x - c)^k} \right| = \lim_{k \to \infty} \left| \frac{a_{k+1}}{a_k} \right| |(x - c)| < 1$$

then the series will converge absolutely for all x satisfying

$$|x - c| < 1/ \lim_{k \to \infty} \left| \frac{a_{k+1}}{a_k} \right|$$

Thus the radius of convergence is at least $1/ \lim_{k \to \infty} |\frac{a_{k+1}}{a_k}|$. Note: if either of the two limits is zero, then the radius of convergence is infinity and the power series converges absolutely for all real numbers. Similarly if either of the two limits is infinity, then the radius of convergence is zero and the series converges only for $x = c$.

Either test can be applied. If a_k involves factorials, then the ratio test is usually best. If a_k involves k in the exponent, then the root test may be best.

Example 1

(a) Consider the power series

$$\sum_{k=0}^{\infty} k!(x - c)^k$$

Since

```
> limit((k+1)!/k!, k=infinity);
```

is infinity, the radius of convergence is zero, and thus the series converges only at $x = c$.

(b) Consider the power series

$$\sum_{k=0}^{\infty} \frac{1}{k!}(x - c)^k$$

Now

```
> limit(k!/(k+1)!, k=infinity);
```

is zero, and so the series converges for all x. Thus we have examples of the extreme cases, where $R = 0$ and $R = \infty$.   △

Example 2

Consider the series

$$\sum_{k=1}^{\infty} \left(1 + \frac{1}{k}\right)^{k^2} (x - c)^k$$

Notice that since convergence depends on the tail of the series, we can start at any subscript. In this case, we start with $k = 1$. Applying the root test (and using our CAS) we have

```
> a := proc(k) (1+1/k)^(k^2) end;
> limit( a(k)^(1/k), k=infinity);;
```

which returns e. Thus the radius of convergence is $R = 1/e \approx 0.3678794412$. $\triangle$

Exercises

1. Conjecture the radius of convergence of the series in part a and b by drawing a multiplot of the first 5 partial sums. (It may be necessary to readjust the horizontal and vertical scales two or three times before obtaining a suitable picture.)

   (a) $\displaystyle\sum_{k=0}^{\infty}(2x)^k$

   (b) $\displaystyle\sum_{k=0}^{\infty}\left(\frac{x}{3}\right)^k$

   (c) From the results of parts (a) and (b), conjecture the radius of convergence of $\displaystyle\sum_{k=0}^{\infty}(cx)^k$. (You may want to look at other specific series similar to those in parts (a) and (b) such as: $\displaystyle\sum_{k=0}^{\infty}\left(\frac{2x}{3}\right)^k$, $\displaystyle\sum_{k=0}^{\infty}(4x)^k$, etc..)

   (d) Give an analytical proof of your conjecture in part c.

2. Conjecture the radius of convergence of the series in part (a) and (b) by drawing a multiplot of the first 5 partial sums. (It may be necessary to readjust the horizontal and vertical scales two or three times before obtaining a suitable picture.)

   (a) $\displaystyle\sum_{k=1}^{\infty}k(2x)^{k-1}\,k$

   (b) $\displaystyle\sum_{k=1}^{\infty}k(\frac{x}{3})^{k-1}\,k$

   (c) From the results of parts (a) and (b), conjecture the radius of convergence of $\displaystyle\sum_{k=1}^{\infty}k(cx)^{k-1}$. (You may want to look at other specific series similar to those in parts (a) and (b) such as: $\displaystyle\sum_{k=1}^{\infty}k(\frac{2x}{3})^{k-1}$, $\displaystyle\sum_{k=1}^{\infty}k(4x)^{k-1}$, etc.)

(d) Give an analytical proof of your conjecture in part c.

3. Conjecture the radius of convergence of the series in part (a) and (b) by drawing a multiplot of the first 5 partial sums. (It may be necessary to readjust the horizontal and vertical scales two or three times before obtaining a suitable picture.)

(a) $\displaystyle\sum_{k=0}^{\infty} \frac{(2x)^{k+1}}{k+1}$

(b) $\displaystyle\sum_{k=0}^{\infty} \frac{(x/3)^{k+1}}{k+1}$

(c) From the results of parts (a) and (b), conjecture the radius of convergence of $\displaystyle\sum_{k=0}^{\infty} \frac{(cx)^{k+1}}{k+1}$. (You may want to look at other specific series similar to those in parts (a) and (b), such as: $\displaystyle\sum_{k=0}^{\infty} \frac{(2x/3)^{k+1}}{k+1}$, $\displaystyle\sum_{k=0}^{\infty} \frac{(4x)^{k+1}}{k+1}$, etc.)

(d) Give an analytical proof of your conjecture in part (c).

4. Find the radius of convergence of the power series $\displaystyle\sum_{k=0}^{\infty} k^4 (5x)^k$.

5. Find the radius of convergence of the power series $\displaystyle\sum_{k=1}^{\infty} k^k (x-2)^k$.

6. Find the radius of convergence of the power series $\displaystyle\sum_{k=0}^{\infty} \frac{(3k)!}{2^k} x^k$.

7. Find the radius of convergence of the power series

$$\sum_{k=0}^{\infty} \frac{(k+5)!}{k!(k+12)!} (x-4)^k$$

8. Find the radius of convergence of the power series $\displaystyle\sum_{k=0}^{\infty} \frac{(k!)^{5/3}}{(3k)!} x^k$.

9. Find the radius of convergence of the power series $\displaystyle\sum_{k=1}^{\infty} \frac{\log(k)}{k} x^k$.

10. Make up a series whose radius of convergence is

(a) 3

(b) infinity

(c) 0

11. Make up a series whose interval of convergence is

(a) (2,8]

(b) the whole real line

(c) [0,0]

Chapter 6

Functions of Several Variables

6.1 Multivariable Differentiation

Prerequisites

- Parametric functions

- Multivariable functions

- The Gradient

- The Chain Rule

Discussion

The lengthy computations that are usually involved in multivariable differentiation often obscure the underlying process and leave the student with a feeling of much more complexity than actually exists. Using a CAS to do the algebra allows the student to concentrate on the essential process involved.

Example 1

Let $f : \Re^2 \to \Re$ be defined by $f(x, y) = (1 + x)^{3/4} y^{2/3}$ where $x(t) = t^2$ and $y(t) = t^4$. We will find the derivative of f with respect to t, $\dfrac{df(x(t), y(t))}{dt}$.

Substitution Method

Since x and y are both functions of t, substituting for x and y yields a function of one variable $g : \Re \to \Re$ and $\dfrac{df(x(t), y(t))}{dt} = \dfrac{dg(t)}{dt(t)}$.

First we define the function:

```
> f:= proc(x,y) (1+x)^(3/4) * y^(2/3) end;
```

Now define x and y as functions of t:

```
> x:= proc(t) t^2 end;
> x(t);
```

$$t^2$$

```
> y:= proc(t) t^4 end;
> y(t);
```

$$t^4$$

Define g by substituting for x and y in the function f:

```
> g:= proc(t) f(x(t),y(t)) end;
```

Differentiate g with respect to t:

```
> gt:= diff(g(t),t);
```

$$gt := \frac{3}{2} \frac{t^{11/3}}{(1+t^2)^{1/4}} + \frac{8}{3}(1+t^2)^{3/4} t^{5/3}$$

Chain Rule Method

First define the parametric function $h : \Re \to \Re^2$ by $h(t) = (x(t), y(t))$. Now composing h with f yields a function $F : \Re \to \Re$. We now differentiate F using the multivariable chain rule. Recall that

$$\frac{dF}{dt} = \frac{\partial f}{\partial x}\frac{dx}{dt} + \frac{\partial f}{\partial y}\frac{dy}{dt}$$

We now turn the computations over to our CAS. Define functions f, x, and y:

```
> f:= proc(x,y) (1+x)^(3/4) * y^(2/3) end:
> f(x,y);
```

$$(1 + x)^{3/4} y^{2/3}$$

```
> x:= proc(t) t^2 end;
> x(t);
```

$$t^2$$

```
> y:= proc(t) t^4 end;
> y(t);
```

$$t^4$$

Let $f1$ be the partial derivative of f with respect to x:

> f1:= diff(f(x,y),x);

$$f1 := \frac{3}{4}\frac{y^{2/3}}{(1+x)^{1/4}}$$

Let $f2$ be the partial derivative of f with respect to y:

> f2:= diff(f(x,y),y);

$$f2 := \frac{2}{3}\frac{(1+x)^{3/4}}{y^{1/3}}$$

Let $h1$ be the derivative of x with respect to t:

> h1:= diff(x(t),t);

$$h1 := 2t$$

Let $h2$ be the derivative of y with respect to t:

> h2:= diff(y(t),t);

$$h2 := 4t^3$$

Now $\dfrac{dF}{dt} = (f1)(h1) + (f2)(h2)$:

> dF:= f1*h1 + f2*h2;

$$dF := \frac{3}{2}\frac{y^{2/3}t}{(1+x)^{1/4}} + \frac{8}{3}\frac{(1+x)^{3/4}t^3}{y^{1/3}}$$

We need to express the answer entirely in terms of t, so we eliminate the terms in x and y by substituting for them:

> subs({x=x(t),y=y(t)},dF);

$$dF := \frac{3}{2}\frac{t^{11/3}}{(1+t^2)^{1/4}} + \frac{8}{3}(1+t^2)^{3/4}t^{5/3}$$

Thus we have the same result by both methods.   △

Example 2

If we let $z = f(x, y) = (1+x)^{3/4} y^{2/3}$, then the function $k(t) = (x(t), y(t), z(t))$, where $x(t) = t^2$ and $y(t) = t^4$, is the parametric equation for a curve C in the surface $z = f(x, y)$ in $\Re^3$. We will find the equation of the line tangent to C at the point $(x(1), y(1), z(1))$. Recall the "point-slope" form of a line in higher dimensional space is expressed parametrically as

$$T(s) = \text{fixed vector to point} + s(\text{direction vector}).$$

In this example the fixed vector is $(x(1), y(1), z(1)) = (1, 1, 2^{3/4})$ and the direction vector is $\dfrac{d(x(t), y(t), z(t))}{dt}$ evaluated at $t = 1$. We first define the functions $x = x(t)$, $y = y(t)$, and $z = z(t)$:

```
> x:= proc(t) t^2 end;
> y:= proc(t) t^4 end;
> z:= proc(t) (1+x(t))^(3/4)*y(t)^(2/3) end;
```

We next compute the components of $T(s) = (T1(s), T2(s), T3(s))$:

```
> T1:= x(1) + s*subs(t=1,diff(x(t),t));
> T2:= y(1) + s*subs(t=1,diff(y(t),t));
> T3:= z(1) + s*subs(t=1,diff(z(t),t));
```

which returns:

$$T1 \quad := \quad 1 + 2s$$
$$T2 \quad := \quad 1 + 4s$$
$$T3 \quad := \quad 2^{3/4} + s\left(\frac{3}{2}\frac{1}{2^{1/4}} + \frac{8}{3}2^{3/4}\right)$$

We can obtain a numerical approximation to $T3$ by applying **evalf**:

```
> evalf(T3);
```

This returns $1.68 + 5.75s$.   △

Exercises

1. Define

$$
\begin{aligned}
f(x, y) &= x^3 y^2 \sin(xy) \\
x(t) &= \sin(t^2 e^t) \\
y(t) &= \sin^2(t \cos(t))
\end{aligned}
$$

(a) Find the derivative of $f(x(t), y(t))$ with respect to t.

(b) Find the equation of the tangent line at $t = 1$ to the curve which is the range of $F(t) = (x(t), y(t), f(x(t), y(t)))$.

(c) What is the relation between the curve in (b) and the surface defined by $z = f(x, y)$?

2. Define

$$
\begin{aligned}
f(x, y, z) &= x^3 y \sin(xyz) e^x \\
x(t) &= \cos(t + t^2 e^t) \\
y(t) &= \sin(t) + e^{\cos(t)} \\
z(t) &= t e^{t^2}
\end{aligned}
$$

(a) Find the derivative of $f(x(t), y(t), z(t))$ with respect to t.

(b) Find the equation of the tangent line to the curve which is the range of $F(t) = (x(t), y(t), z(t), f(x(t), y(t), z(t)))$.

(c) What is the dimension of the set of all points satisfying $w = f(x, y, z)$?

3. Let $f(x, y, z) = 2xy^2z^3$, $x(t) = 3t^2$, $y(t) = \sin(t)$, and $z(t) = \cos(t)$. Find the second derivative of $f(x(t), y(t), z(t))$ with respect to t by using

(a) the substitution method

(b) the Chain Rule

6.2   Extrema in Several Variables

Prerequisites

- Some knowledge of extrema in several variables

- The section on Extremal Values

Discussion

The problem of finding extrema for functions of several variables is similar to that of finding extrema for functions of one variable (i.e., determine the critical points and then check them for extrema). In this section we will find (unconstrained) extrema for functions of several variables, using our CAS to carry out the often tedious computations involved.

We will assume that f has continuous second order partial derivatives. If an extrema occurs at a point p on the interior of the domain of definition of f, then the partial derivatives of f are all zero. Thus to find extrema of f we must:

1. Find the critical points of f, i.e., those points at which all the partial derivatives vanish. To do this, we must find all solutions to the set of equations $\{\dfrac{\partial f}{\partial x} = 0 : x$ a variable$\}$.

2. Check the critical points for extrema. To do this, we can consider the definition or graph of the function, or we may be able to use the multivariate equivalent of the second derivative test. For functions of two variables, x and y, this test is the following. Assume p is a critical point and define the *discriminant* D at p by

$$D = \frac{\partial^2 f(p)}{\partial x^2} \frac{\partial^2 f(p)}{\partial y^2} - \left(\frac{\partial^2 f(p)}{\partial x \partial y} \right)^2$$

(a) If $D > 0$ and $\dfrac{\partial^2 f(p)}{\partial x^2} > 0$, then f has a relative minimum at p.

(b) If $D > 0$ and $\dfrac{\partial^2 f(p)}{\partial x^2} < 0$, then f has a relative maximum at p.

(c) If $D < 0$, then f has neither a relative maximum nor a relative minimum at p. In this case, p is called a *saddle point*.

(d) If $D = 0$, anything can happen, and the test fails.

3. Finally we need to examine the boundary of the domain of definition of f for extrema.

Example 1

Our first example will show that if the discriminant is zero, then anything can happen. Consider the following three functions defined on all of the plane:

$$\begin{aligned}
f(x, y) &= x^4 + y^4 \\
g(x, y) &= -(x^4 + y^4) \\
h(x, y) &= x^3 + y^3
\end{aligned}$$

For each of these three functions, the functions, the first partial derivatives, and the second partial derivatives all vanish at $(x, y) = (0, 0)$. So the discriminant $D = 0$. However, f has a minimum at $(0, 0)$ (since it is positive for any other point), g has a maximum at $(0, 0)$ (since it is negative for any other point), and h has a saddle point at $(0, 0)$ (since the function value increases as x increase and decreases as x decreases). Thus if the discriminant is zero we cannot conclude anything.   △

Example 2

We will find the extrema of $f(x, y) = (x - 2)^2 y + y^2 - y$ on the triangle bounded by the lines $x = 0$, $y = 0$, and $x + y = 4$.

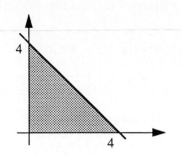

We define f, compute the partial derivatives, and solve the set of equations for the critical points:

```
> f:= proc(x,y) (x-2)^2*y+y^2-y end;
> fx:= diff(f(x,y),x);
> fy:= diff(f(x,y),y);
> solve({fx=0,fy=0},{x,y});
```

The **solve** command returns

$$\{x = 2, y = 1/2\}, \quad \{x = 1, y = 0\}, \quad \{y = 0, x = 3\}$$

as the critical points of f on the entire plane. Only $(2, 1/2)$ is in the interior of the triangle. We compute the discriminant:

```
> fxx:= diff(fx,x);
> fyy:= diff(fy,y);
> fxy:= diff(fx,y);
> subs({x=2,y=1/2}, fxx*fyy - fxy^2);
```

which returns the value of 2. Since

```
> subs({x=2,y=1/2},fxx);
```

returns the value of 1, f has a local minimum at $(2, 1/2)$ of $f(2, 1/2) = -1/4$.

Now we check f on the boundary of the triangle. We check each boundary line separately.

On the line from $(0,0)$ to $(4,0)$ we have $y = 0$ and f restricted to this line is $f_1(x) = f(x, 0) = 0$. f_1 is constant, so each point is a local extremum, with value 0.

On the line from $(0,0)$ to $(0,4)$ we have $x = 0$ and f restricted to this line is $f_2(y) = f(0, y) = y^2 + 3y$. The problem is now reduced to finding the extrema of a function of a single variable, f_2. On the interior of the line segment, $0 < y < 1$, $f_2'(y) = 2y + 3 > 0$ so f_2 is strictly increasing. Thus f_2 has a minimum at $(0,0)$ of 0 and a maximum at $(0,4)$ of $f_2(4) = 28$.

On the third part of the boundary, the line between $(0, 4)$ and $(4, 0)$, we have $x + y = 4$ or $y = 4 - x$. Restricting f to this line gives $f_3(x) = f(x, 4 - x)$. Applying our CAS,

```
> solve(diff(f(x,4-x),x) = 0,x);
```

we have only one critical point at $x = 3$ ($y = 4 - 3 = 1$). At this point

```
> subs(x=3,diff(f(x,4-x),x$2));
```

returns 0, so the second derivative test fails for f_3. But we can always plot:

```
> plot(f(x,4-x),x=0..4);
```

which returns

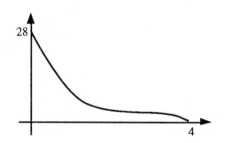

We see that f_3 only has extrema at the endpoints, which we have already found when examining the other portions of the boundary.

Combining our results, we see that on the triangle f has a global maximum of 28 occurring at $(0, 4)$ and a global minimum of $-1/4$ occurring at $(2, 1/2)$. $\triangle$

Example 3

We will find the extrema of $f(x, y, z) = x^2(y + z^2) + x\,y + z$ on all of $\Re^3$. We define the function and solve for the critical points:

```
> f:=proc(x,y,z) x^2*(y+z^2)+x*y+z end;
> fx:= diff(f(x,y,z),x);
> fy:= diff(f(x,y,z),y);
> fz:= diff(f(x,y,z),z);
> solve({fx=0,fy=0,fz=0},{x,y,z});
```

returns

$$\{y = -1/2, x = -1, z = -1/2\}$$

so there is only one critical point. How can we determine if f has an extremum or a saddle point at this point? There is a second derivative test for three variables, similar to the test for two variables, but it is rather complex. We might try to analyze the function from its definition, but instead we will plot f. If f has an extremum at the point, then every cross section (with all but one variable fixed) will show the same kind of extremum at the point. So we plot:

```
> plot(f(x,-1/2,-1/2),x=-2..0);
> plot(f(-1,y,-1/2),y=-1..0);
> plot(f(-1,-1/2,z),z=-1..0);
```

returns

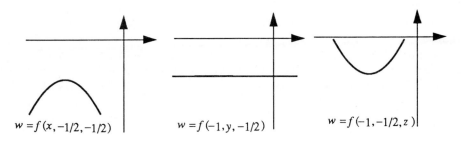

$$w = f(x, -1/2, -1/2) \qquad w = f(-1, y, -1/2) \qquad w = f(-1, -1/2, z)$$

From the plot on the left with only x varying, we see that if f has an extremum, it must be a maximum. The second plot, with y varying, does not contradict this. However, the third plot, with z varying, allows only a minimum as extremum. Thus f must have a saddle point at $(-1, -1/2, -1/2)$, and no extrema in $\Re^3$.

Warning: even if the three plots above all showed a maximum, f need not have a maximum at the point, since f could increase along some non-axis direction.   △

Example 4

We will find the extrema of $f(x, y, z) = x(x - 3y) + y^2 + z^2$ on all of $\Re^3$. We proceed exactly as in the last example, obtaining exactly one critical point, $(0, 0, 0)$. When we plot the graphs, as before, we obtain plots consistent with a minimum at the origin.

We could plot some more graphs, going along other directions. For example, to plot along a line midway between x and y, we would plot $f(x, x, 0)$.

Another approach is to sample the values of f near the origin. If f has a minimum at the point, then all values near it should be non-negative. Thus we sample points near $(0, 0, 0)$ with the aid of our CAS, printing out any at which f is negative. We will sample points on a 20 by 20 by 20 grid centered at the origin, using nested for loops:

```
> f:= proc(x,y,z) x*(x - 3*y) +y^2+z^2;
≫ for i from 1 to 20 do
≫         for j from 1 to 20 do
```

```
≫             for k from 1 to 20 do
≫                 x:= -0.1+i*(0.01);
≫                 y:= -0.1+j*(0.01);
≫                 z:= -0.1+k*(0.01);
≫                 v:= f(x,y,z);
≫                 if v < 0 then print(x,y,z,v) fi;
≫             od;
≫         od;
≫ od;
```

This soon begins to print negative values; thus f has a saddle point at the origin rather than a minimum.

Can you find a plot that also shows that f does not have a minimum at the origin?

Warning: neither the sampling method nor graphing (which is really a form of sampling) can guarantee results – only analytic techniques can do this. However, we would be very surprised if extensive sampling gave an incorrect result.   △

Example 5

In this example we will consider a case where the exact solutions to the system of equations cannot be found. We will find the extrema of $f(x, y, z) = x^4 + y^2 + z^2 + xy - yz + 3x$. We proceed as before: define f, compute the partial derivatives f_x, f_y, and f_z, and solve the system of equations. This time the Maple CAS returned the message:

$$\begin{aligned} \{y &= \quad \text{RootOf}(27Z^3 - 2Z - 6) \\ x &= \quad -3/2 \, \text{RootOf}(27Z^3 - 2Z - 6) \\ z &= \quad 1/2 \, \text{RootOf}(27Z^3 - 2Z - 6)\} \end{aligned}$$

This means that Maple was not able to find an exact root of the polynomial $27Z^3 - 2Z - 6$. We can use the Bisection Algorithm or Maple's built-in numerical root finder to obtain an approximate root:

```
> rt:= fsolve(27*Z^3 -2*Z - 6 = 0, Z);
```

This returns $rt := .646413971201$. (We can plot $p = 27 * Z^3 - 2Z - 6$ to check that $rt := .646413971201$ is the only root.)

We can now define x, y, z in terms of this approximate value:

```
> xr:= rt; yr:= -3/2*rt; zr:= 1/2*rt;
```

This returns

$$\begin{aligned} xr &:= \quad -.969620956803 \\ yr &:= \quad .646413971201 \\ zr &:= \quad .323206985601 \end{aligned}$$

as the approximate location of a critical point. As a check, we evaluate the partial derivatives at this point:

```
> subs({x=xr,y=yr,z=zr},{fx,fy,fz});
```

returns $\{.1 * 10^{-11}, -.2 * 10^{-10}, -.4 * 10^{-11}\}$, which is close to zero.

Now, is this critical point an extremum? From the definition of the function we see that the function values increase when (x, y, z) is far from the origin, so we expect a minimum. Since we found only one critical point, we hope it is a minimum. We plot the function along the coordinate directions as before:

```
> plot(f(x,yr,zr),x=xr-1..xr+1);
> plot(f(xr,y,zr),y=yr-1..yr+1);
> plot(f(xr,yr,z),z=zr-1..zr+1);
```

All three of the plots show a minimum at the point (xr, yr, zr) of approximately $f(xr, yr, zr) = -2.33834128612$. Sampling about (xr, yr, zr) with **for** loops shows no descent below this value, so we conclude that f has a global minimum of about -2.33834128612 at about $(-.969620956803, .646413971201, .323206985601)$.
$\triangle$

Exercises

1. Give examples, different from Example 1, where the discriminant is zero and the point is a maximum, a minimum, or a saddle point.

2. Find the extrema of $f(x, y) = x^2 + 2xy + y^2$ on the disk of radius 2 about the origin, i.e., all (x, y) with $x^2 + y^2 \le 4$.

3. Find the extrema of $f(x, y) = e^{xy} \sin(x)$ on the square of side 2 centered on the origin, i.e., all (x, y) with $|x| \le 1$ and $|y| \le 1$.

4. Find the extrema of $f(x, y) = \arctan(xy)$ on the unit square with lower left corner at the origin, i.e., all (x, y) with $0 \le x \le 1$ and $0 \le y \le 1$.

5. Find the extrema of $f(x, y) = (x^2 + y^2) \log(x + y)$ on the annulus centered at the origin with inner radius $\sqrt{2}$ and outer radius 2, i.e., all (x, y) with $2 \le x^2 + y^2 \le 4$.

6. Find the extrema of $f(x, y, z) = x(y + z) + y^2 + z^2$.

7. Find the extrema of $f(x, y, z) = x^2 + y^2 + z^2 + x^2y$.

8. Find the extrema of $f(x, y, z) = 3x^2 + y^4 + 4z^2 - 2xy - yz + 3x$.

9. Find the extrema of $f(x, y, z) = x^2 + y^2 + z^3 + 2x$.

10. Find the extrema of $f(x, y, z) = x^2 + y^4 + z^2 - 3xz + 2x$.

11. Suppose an experiment is performed and a set of data points $\{(x_i, y_i)\}$ is obtained. The experimenter may want to draw a straight line "through" these points for predictive purposes. If the points do not lie on a straight line, the experimenter wants to find a line that "fits" the data as closely as possible. A standard procedure for doing this is the following *Method of Least Squares.*

The objective is to determine a line, $y = mx + b$, for which the sum of the squares of the differences between the observed value y_i and the predicted (y value) $mx_i + b$ is a minimum. That is, we need to determine the parameters m and b that will minimize the error function defined by

$$E(m, b) = \sum_{i=1}^{n} (y_i - mx_i - b)^2$$

(a) Find the equation of the straight line that best fits (in the sense of the Method of Least Squares) the data $\{(-2, -3), (-1, 0), (1, 1), (2, 3), (4, 3)\}$. Plot your line along with the data points.

(b) Show that if the data is $\{(x_1, y_1), (x_2, y_2)\}$, the Method of Least Squares gives the line passing through these two points.

6.3   Lagrange Multipliers

Prerequisites

- An introduction to the use of Lagrange multipliers

Discussion

Lagrange multiplier problems are usually difficult and time consuming because they involve setting up the problem; performing differentiations to obtain a system of equations; solving the system of equations for potential extrema; and examining points for extrema. The major difficulties occur in the third step since, in general, the system of equations is non-linear; even solving a linear system is time consuming. This means that only a few examples can be given, carefully selected so that the resulting system of equations is easy to solve. With a CAS, students can focus on the essential steps in the process.

Example

Suppose we want to find the extrema of the function

$$f(x, y, z, t) = x^2 + 2y^2 + z^2 + t^2$$

subject to the conditions

$$x + 3y - z + t = 2 \quad \text{and} \quad 2x - y + z + 2t = 4$$

From geometric considerations, we expect a single minimum since the conditions define a plane in $\Re^4$ and the level curves of f are ellipses in the plane. Using a CAS, we would define the expressions:

```
> f:= x^2 + 2*y^2 + z^2 + t^2;
> g1:= x + 3*y - z + t - 2 ;
> g2:= 2*x - y + z + 2*t - 4;
```

The steps outlined above are carried out in the following manner.

1. Set up the problem:

```
>   h:= f + lambda1*g1 + lambda2*g2;
```

2. For each variable, compute the partial derivative:

```
> expn1:= diff(h,x);
> expn2:= diff(h,y);
> expn3:= diff(h,z);
> expn4:= diff(h,t);
> expn5:= diff(h,lambda1);
> expn6:= diff(h,lambda2);
```

3. Solve the six linear equations in six unknowns:

```
> solve({expn1=0,expn2=0,expn3=0,expn4=0,expn5=0,
> expn6=0},{x,y,z,t,lambda1,lambda2});
```

The CAS returns the exact solution(s) in a list:

$$\{x = 67/69, \ y = 2/23, \ z = 14/69, \ t = 67/69,$$

$$lambda1 = -26/69, \ lambda2 = 18/23\}$$

4. Evaluate f at the point obtained in step 3:

```
> subs({x=67/69,y = 2/23,z=14/69,t=67/69 },f);
```

which returns 134/69. Thus, we have the minimum value of f subject to the given constraints.   △

Exercises

1. Explore the problem: A rod one meter in length is to be cut into n pieces. Determine the lengths of the pieces if the product of their lengths is to be a maximum.

   (a) Answer the problem when $n = 2$. (This does not require using Lagrange multipliers.)

   (b) Answer the problem when $n = 3$.

   (c) Answer the problem when $n = 4$.

   (d) Conjecture an answer to the problem based on the special cases, $n = 2, 3$, and 4. (It may be necessary to look at additional special cases.) Give a geometrical argument supporting your conjecture.

   (e) Prove your conjecture. Or, disprove your conjecture and return to part (d).

2. Find the minimum distance from the surface $x^2 - y^2 + z^2 = 3$ to the origin.

3. If a, b, c, and d are positive numbers, find the maximum value that $f(x, y, z, w) = ax + by + cz + dw$ can assume on the four dimensional unit sphere, $x^2 + y^2 + z^2 + w^2 = 1$.

4. Find the minimum of the function

$$f(x, y, z, t) \; = \; x^2 + y^2 + z^2 + t^2$$

   subject to the conditions

$$x + y - z + 2t \; = \; 2 \ \text{ and } \ 2x - y + z + 3t \; = \; 3$$

5. Find the minimum of the function

$$f(x, y, z, t) \; = \; 2x^2 + y^2 + z^2 + 2t^2$$

   subject to the conditions

$$x + y + z - t \; = \; 1, \ \ 2x + y - z + 2t \; = \; 2, \ \text{ and } \ x - y + z - t \; = \; 4$$

6 Find the extremal values of $f(x, y, z) = xyz$ where the domain of f is the set of points lying on the intersection of the planes $x + y + z = 40$ and $z = x + y$. State and verify the nature of the extremal values, i.e., maximum or minimum.

6.4   Taylor's Polynomials in Several Variables

Prerequisites

- The section on Taylor's Theorem

Discussion

Taylor's Theorem allows us to approximate functions of a single variable by a polynomial. Taylor's polynomials for several variables allow us to approximate functions of several variables by polynomials in several variables. As with functions of a single variable, an approximating Taylor polynomial of degree n will be the best approximating polynomial of degree less than or equal to n. In this section we find Taylor polynomials for functions of several variables and look at some applications.

We can compute Taylor's polynomial for a function of several variables by computing the Taylor's polynomial for the related functions of a single variable, and then doing the necessary operations or substitutions. A few examples will make this rather vague recipe clearer.

Example 1

We will find the Taylor's polynomial for $f(x, y) = \sin(x) e^y$ of degree 5 when we expand about $(0,0)$. First we find the Taylor's polynomial expansions for the individual functions of a single variable:

```
> sinp:= convert(taylor(sin(x),x=0,6),polynom);
> expp:= convert(taylor(exp(y),y=0,6),polynom);
```

(Recall that in Maple the third parameter, 6, is one more than the desired degree.) This returns

$$\text{sinp} := \quad x - \tfrac{1}{6} x^3 + \tfrac{1}{120} y^5$$
$$\text{expp} := \quad 1 + y + \tfrac{1}{2} y^2 + \tfrac{1}{6} y^3 + \tfrac{1}{24} y^4 + \tfrac{1}{120} y^5$$

We now expand the product:

```
> expand(sinp*expp);
```

which returns

$$x \;+\; xy + \frac{1}{2} xy^2 + \frac{1}{6} xy^3 + \frac{1}{24} xy^4 + \frac{1}{120} xy^5 - \frac{1}{6} x^3 - \frac{1}{6} x^3 y$$
$$-\; \frac{1}{12} x^3 y^2 - \frac{1}{36} x^3 y^3 - \frac{1}{144} x^3 y^4 - \frac{1}{720} x^3 y^5 + \frac{1}{120} x^5$$
$$+\; \frac{1}{120} x^5 y + \frac{1}{240} x^5 y^2 + \frac{1}{720} x^5 y^3 + \frac{1}{2880} x^5 y^4 + \frac{1}{14400} x^5 y^5$$

Keeping only the terms of degree less than or equal to 5 and rearranging by increasing degree, we have the best approximating polynomial to $f(x, y)$ at $(0, 0)$ of degree less than or equal to 5:

$$x \;+\; xy - \frac{1}{6} x^3 + \frac{1}{2} xy^2 + \frac{1}{6} xy^3 - \frac{1}{6} x^3 y + \frac{1}{24} xy^4$$
$$-\; \frac{1}{12} x^3 y^2 + \frac{1}{120} x^5$$

(Some CASs, such as MACSYMA, will compute this with one command.)   △

Example 2

We will find the best approximating polynomial to $f(x, y) = \cos(x + y)$ at $(0, 0)$ of degree less than or equal to 5 . In this case we will compute the Taylor's polynomial for $\cos(z)$ at $0 + 0 = 0$ and then substitute $z = x + y$.

```
> subs(z=x+y,convert(taylor(cos(z),z=0,6),polynom));
```

returns

$$1 - 1/2\,(x + y)^2 + 1/24\,(x + y)^4$$

as the best approximating polynomial to f at $(0, 0)$ of degree less than or equal to 5.   △

Example 3

We will find the best approximating polynomial to $f(x, y, z) = x\sin(y\,z) + \log(x + y)$ at $(0, 1, 0)$ of degree less than or equal to 6. (We choose $(0, 1, 0)$ since we don't want to evaluate log at 0). Using our CAS, we find the Taylor's polynomials of the functions of a single variable (expanding sine at $1*0 = 0$ and log at $0 + 1 = 1$), make the substitutions, and form the desired combination:

```
> sinp:= subs(w=y*z,convert(taylor(sin(w),w=0,7),polynom));
> logp:= subs(w=x+y,convert(taylor(log(w),w=1,7),polynom));
> expand(x*sinp+logp);
```

Keeping only the terms of degree less than or equal to 6 we have

$$
\begin{aligned}
-\frac{49}{20} \;&+\; 6x + 6y - \frac{15}{2}\,(x^2 + y^2) - 15xy + \frac{20}{3}\,(x^3 + y^3) + xyz \\
&+\; 20(x^2y + xy^2) - \frac{15}{4}(x^4 + y^4) - 15(x^3y + xy^3) - \frac{45}{2}\,x^2y^2 \\
&+\; 6(x^4y + xy^4) + \frac{6}{5}\,(x^5 + y^5) + 12(x^3y^2 + x^2y^3) - (x^5y + xy^5) \\
&-\; \frac{5}{2}\,(x^2y^4 + x^4y^2) - \frac{10}{3}\,x^3y^3 - \frac{1}{6}\,(x^6 + y^6)
\end{aligned}
$$

as the approximating polynomial.   △

Discussion: Applications

Consider a differentiable function f of one variable. Its Taylor's polynomial of degree one is $f(x_0) + f'(x)(x - x_0)$. This is the best approximating polynomial (of degree one) to f at x_0, since the line $y = f(x_0) + f'(x)(x - x_0)$ is the tangent line to the curve $y = f(x)$ at the point $(x_0, f(x_0))$.

The same situation holds for functions of several variables. Let f be a function of two variables, and consider the surface $z = f(x, y)$. The tangent plane to the surface at the point $(x_0, y_0, f(x_0, y_0))$ will be given by the best first degree approximating polynomial to f at (x_0, y_0), the Taylor's polynomial of degree one expanded about (x_0, y_0):

$$z = f(x_0, y_0) + \frac{\partial f}{\partial x}(x_0, y_0)(x - x_0) + \frac{\partial f}{\partial y}(x_0, y_0)(y - y_0)$$

(Note that this equation is the same one obtained from using the fact that the gradient ∇f is perpendicular to the tangent plane at (x_0, y_0).)

Example 4

We will find the tangent plane to the surface $z = \sin(4x+y)$ above $(x, y) = (1, 2)$. We find the first degree Taylor's polynomial for sin expanded about $4 * 1 + 2 = 6$ and substitute:

```
> z:= subs(w=4*x+y, convert(taylor(sin(w),w=6,2),polynom));
```

This returns

$$z := \sin(6) + \cos(6)(4x + y - 6)$$

as the tangent plane.   △

Example 5

We will find the best quadratic (second degree) approximation to the surface $z = e^x \tan(y)$ at $(0, 0)$. We want the Taylor's polynomial for $f(x, y) = e^x \tan(y)$ of degree 2 at $(0, 0)$. Using our CAS,

```
> z:= convert(taylor(exp(x),x=0,3),polynom)
>        *convert(taylor(tan(y),y=0,3),polynom);
```

returns

$$z := (1 + x + 1/2 \, x^2)y$$

Keeping only the terms of degree less than or equal to two gives

$$z = (1 + x)y$$

as the equation of the approximating quadratic.   △

Exercises

1. Find the best 6th degree polynomial approximation to $f(x, y) = \tan(\sin(x + y))$ at $(0, 0)$.

2. Find the best 6th degree polynomial approximation to $f(x, y) = \sin(x) \cos(x + y)$ at $(0, 0)$.

3. Find the best 3rd degree polynomial approximation to
   $f(x, y) = \log(x)\tan(x\,y)$ at $(1, 0)$.

4. Find the best 5th degree polynomial approximation to
   $f(x, y) = \log(x + 3y)\sin(x)$ at $(0, 1)$.

5. Find the best 4th degree polynomial approximation to
   $f(x, y, z) = \sin(x + y\,z)\cos(x + y)$ at $(0, 0, 0)$.

6. Find the equation of the tangent plane to the surface $z = \sin(x - 2y)$ above
   $(0, 0)$.

7. Find the equation of the tangent plane to the surface $z = e^x \log(y + 2x)$
   above $(1, 2)$.

8. Find the equation of the best approximating quadratic to the surface
   $z = \cos(x + x\,y)\sin(y)$ when $(x, y) = (1, 2)$.

9. Find the equation of the best approximating quadratic to the surface
   $z = x + \tan(e^y)$ above $(1, 0)$.

10. Find the equation of the tangent hyperplane to the hypersurface
    $w = \sin(x\,z - 2y\,x)$ when $(x, y, z) = (1, 0, 0)$.

11. Find the equation of the tangent hyperplane to the hypersurface
    $w = x\,y + \cos(x^2 y\,z)$ when $(x, y, z) = (1, 1, 0)$.

Appendix A

Computer Algebra Systems

A.1 Calculus T/L

Discussion

Calculus T/L is a computer environment for doing, learning, and teaching calculus, and runs on the Macintosh. It incorporates the same kernel as Maple, and it can be used to assist in solving calculus problems (one can either type commands or use T/L's unique context sensitive help system), and to deliver guided explorations of mathematical ideas, and as an environment for writing programs.

Among its capabilities are

- Expansion and factoring of algebraic expressions

- Calculus (differential and integral)

- Solution of equations and systems of equations

- Computation of limits

- Computation of sums

- 2D graphics and simple animation

Calculus T/L is designed primarily to be an affordable aid for students in learning calculus.

Getting Started

Start Calculus T/L by double clicking on the Calculus T/L icon:
Calculus T/L

After a few moments your computer screen will look like this

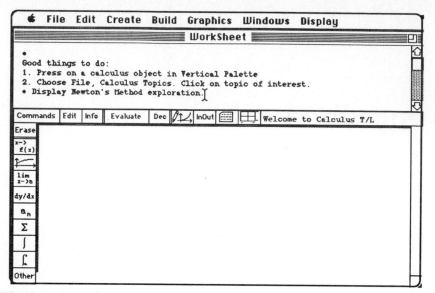

The Calculus T/L screen is divided into two main parts. The bottom part is an electronic White Board where solutions to problems are constructed. The top portion is a window into the Calculus T/L calculus book. This window displays one panel at a time. You may quit Calculus T/L at any time by choosing Quit from the File menu.

Example

Problem solving with Calculus T/L begins by creating one or more mathematical objects and proceeds by applying appropriate commands to obtain complete solutions. In most cases solutions can be generated with minimal typing. In the example below we will create a function and then produce a solution to a definition of a derivative problem that looks like one from a text book.

Create the cubing function by clicking $\boxed{\begin{matrix} x \to \\ f(x) \end{matrix}}$, replacing the predefined function `x*sin(x+2)` by `x∧3`, and clicking $\boxed{\texttt{Define}}$.

Form the difference quotient by choosing `Difference Quotient` from the `Create` menu and clicking `OK`.

Click $\boxed{\text{Commands}}$ to obtain commands that apply to quotient expressions. Choose and click on appropriate commands from the first two panels that appear in the top of the screen. Your solution will be constructed in the White Board and should look like the one below.

$$\boxed{f : x \longrightarrow x^3}$$

$$\frac{f(x+h) - f(x)}{h} \; = \; \frac{(x+h)^3 - x^3}{h} \; = \; \frac{3x^2 + 3xh^2 + h^3}{h}$$

$$= \frac{h(3x^2 + 3xh + h^2)}{h} = 3x^2 + 3xh + h^2$$

$$\lim_{h \to 0} 3x^2 + 3xh + h^2 = 3x^2$$

$\boxed{\text{Commands}}$ provides access to commands that are appropriately applied to each type of object on the White Board. Commands are rarely typed in Calculus T/L. Instead command choices are made available when they are needed. $\triangle$

Guided Explorations

The easiest way to begin using Calculus T/L is to use guided explorations. Explorations allow a student to focus on mathematics with little concern for how to use the computer. Calculus T/L comes with explorations about area, asymptotes, best linear approximations, bisection, composition, curve fitting, definition of derivatives, derivatives of compositions, chain rule motivation, direction of graphs, expression, building, additive property, fixed points, function construction, geometric series, graphical composition, graphing using derivatives, induction, integration, inverse functions, iteration, limits, linear motion, mean-value property, Monte Carlo integration, mean-value theorem proof, Newton's method, graphing one parameter families, polar plotting, function ranges, sequences, series, series and area, step by step derivatives, substitution, Taylor series, unique zeros, and computing zeros.

Gain access to guided explorations by choosing **Calculus Topics** from the **File** menu or by clicking the appropriate button in the vertical palette of the White Board.

A.2   Derive

Discussion

Derive is an interactive symbolic mathematics system which uses a menu interface. It runs on MS-DOS computer systems, requiring Hercules, CGA, EGA, or VGA graphics. Among its capabilities are:

- Arbitrary integer arithmetic

- Calculus (differentiation, integration, vector)

- Expansion and factoring of expressions

- Graphics

- Matrix operations

- Solutions of equations (systems and differential)

- Taylor's polynomials

Our description applies directly to version 1.6, but since only the basic features are discussed, it should also apply to later versions. For more information, see the user's manual for Derive.

Derive runs on MS-DOS and supports a variety of graphics adapters. It has substantial capabilities, especially considering its minimal requirements. Its only major lack is control structures.

Getting Started

To start Derive enter **derive**. (Derive will read an initialization file "DE-RIVE.INI" if you have one.) The Derive screen is divided into a display area, a command area, a status area, and a message line. The display area shows the results of the commands, numbered for reference; the status area shows the current state of Derive; the message line says what to do next (initally it will say "Enter option"); and the command area lists the command options available. Each command has one letter in upper case (usually the first, unless two commands have the same first letter).

Notation: In this discussion, commands selected from the menu or keyboard are in **bold face**. Expressions entered by the user are, as usual, shown in `typewriter` face. In Derive, functions are displayed in upper case and variables are displayed in lower case (no matter what case is used to enter the expression).

A command can be selected by pressing the **Space** bar to highlight the desired command, and then pressing the **Enter** key. A command can also be selected by entering the command's upper case letter.

The commands at this level are:

- **Author**: enter an expression, which will be displayed.

- **Build**: combine existing expressions into a new expression.

- **Calculus**: go to submenu for calculus operations.

- **Declare**: declare domains of variables or functions.

- **Expand**: expand expressions.

- **Factor**: factor expressions.

- **Help**: go to submenu for help on operations.

- **Jump**: select an expression in the display window.

- **soLve**: solve an equation.

- **Manage**: control the application of (trigonometric and other) transforma-tions.

- **Options**: control of display of expressions (e.g., precision).

- **Plot**: plot expressions.

- **Quit**: quit Derive.

- **Remove**: remove expressions from display window.

- **Simplify**: simplify expressions.

- **Transfer**: input and output of expressions to and from external files.

- **moVe**: rearrange expressions in display window.

- **Window**: to create and manage windows.

- **approX**: approximate an expression.

Commands often have subcommands. As an example, to differentiate an expression, the user selects the sequence **C** (for **Calculus**) and then in the submenu selects **D** (for **Differentiate**). The assumed expression to be differentiated and the assumed variable to differentiate with respect to will be listed. If these choices are correct, the user presses the **Enter** key, otherwise the user can modify these default selections. The differentiation command is now applied to the expression, but evaluation need not be done at once. The **Simplify** or **Expand** command can then be applied to obtain the derivative. There are a number of steps, but they are natural and proceed quickly with only a few keystrokes. Commands can also be authored. For example, to do the differentiation, the user would author `DIF(SIN(x),x)` and then **Expand**.

In the list of commands below, we will give the upper level choice of commands. The expression to which the command is applied (if not the default highlighted expression) is first selected with the **Jump** command or by using the arrow keys to move around within the display window. For example, the above differentiation sequence will be abbreviated **C D S** (**Calculus Differentiate Simplify**) where we assume the desired expression has been selected. Commands will be spelled out the first time they occur. No entry is shown if the default value is used. For example, to evaluate a limit at zero, the default evaluation point is zero, so only the commands **C Limit S** (**Calculus Limit Simplify**) are selected; the other options for variable (x) and evaluation point (0) use the default settings.

Some Derive Commands

(1) Assign an expression to a name: **Author** `(x+1)∧3 + 3*x - sin(x)`
    ($\wedge$ is exponentiation)   (it is numbered)
(2) Comment: `A ; text after a semicolon`
(3) Constants: pi, e, i   `pi #e #i`
(4) Define function f (x variable): `A f(x) := x∧2 + 1`
    Forget definition: `A f(x) :=`
(5) Denominator of ratio: **Build** (after selecting expression)

(6) Differentiate: **Calculus Differentiate Simplify**
     Twice: **C D C D S**
     Several variables: **C D x C D y S**

(7) Expand expression: **Expand**

(8) Evaluate to number: **approX**
     To k decimal digits: **Options Precision Digits k X**

(9) Factor expression: **Factor**

(10) **For** loop: Not available

(11) **If** conditional: Not available (try **STEP** function)

     E.g., $f(x) = \begin{cases} 1 & \text{for } x < 1 \\ x & \text{for } x \geq 1 \end{cases}$ **A F(x) :=(x-1)*STEP(x-1)+1**

(12) Help: **Help**

(13) Integrate: **C Integrate S**
     Definite integral (0 to π):   **C I x 0 pi S**
     Numerical approximation: **C I x 0 pi X**

(14) Interrupt computation: Press **ESCAPE-key**

(15) Limit at 0:   **C Limit**
     at infinity: **C L inf**

(16) Numerator of ratio: **Build** (after selecting expression)

(17) Part: second term in expression: Select with arrow keys

(18) Plot: **Plot**
     Several functions   **P P**

(19) Previous expression: Expressions are numbered

(20) Read saved file: **Load** *filename*

(21) Save expression in file: **Transfer Save** *filename*

(22) Simplify an expression: **Simplify**

(23) Solve equation: **soLve**
     Linear system: Not available – use matrix operations
     Numerical: Approximation mode

(24) Stop (end) session: **Quit**

(25) Substitute into expression: **Manage Substitute** *expression*

(26) Sum an expression: **C Sum** *lower-limit upper-limit*

(27) Taylor series (at 0, deg. 5):   **Taylor**
     Several variables: Apply to each variable separately

(28) Time elapsed since start: Time given for each operation

(29) Truncate to an integer: Not available

(30) **While** loop: Not available

Example of Usage

Suppose we have the function $f : \Re \longrightarrow \Re$ defined by

$$f(x) = x^2 + \cos(x)$$

To define this function for Derive, we enter

>   **Author f(x) := x∧2 + cos(x)**

Then we can evaluate the selected f at 0 by

>   **Simplify**

or take the limit as x approaches 1 by selecting f and entering

>   **Limit x 1**

Every expression is numbered and can be referred to by that number. So we can enter

>   **Author x∧2 + COS(x)**

and it will be given a number, **#n**, for reference. This is not a function, but we can still use this expression in, e.g., finding the limit, by selecting it or giving the number in place of the default:

>   **Limit #n x 1**

These commands can also be given by authoring

>   **Author LIM(x∧2 + COS(x),x,1)**

We can substitute 1 for x in expression **#n** by

>   **Manage Substitute #n x 1**

Finally, we can numerically evaluate symbolic expressions like sin(1) by

>   **Author SIN(1) approX**

A.3   MACSYMA

Discussion

MACSYMA is an interactive symbolic mathematics system. It has a wide range of capabilities that enable a user to apply mathematical transformations to symbolic inputs to yield either symbolic results, numeric results, or FORTRAN programs. Some of the problems MACSYMA can solve are:

- Arbitrary precision arithmetic
- Calculus (differential, integral, and vector)
- Expansion and factoring of expressions
- Graphing
- Infinite and finite series, limits, and products

- Matrix operations

- Solutions of equations (systems, non-linear, differential)

- Taylor's polynomials

In addition, there is a large variety of functions for managing the MACSYMA system.

MACSYMA can be used as a programming language to write procedures to solve individual problems. There is a library ("share") of user contributed procedures.

Our description applies directly to version 309.2, but since only the basic features are discussed it should also apply to later versions.

Getting Started

To start MACSYMA, enter `macsyma`. To exit MACSYMA, enter `quit()`. Note that MACSYMA does not distinguish between cases: "x" and "X" will be considered the same character.

MACSYMA has extensive on-line help facilities. There is an on-line primer (enter `primer(help)`) with lessons, files of commented demonstrations of various topics, a description command (enter `describe(`*function-name*`)`), and a keyword command (enter `apropos(`*string*`)`). Since apropos matches parts of keywords, it is best to use the smallest distinguishing part. For example, instead of `apropos(differentiation)`, which may not match, try `apropos(diff)`.

Some MACSYMA Commands

(cn) indicates the prompt from MACSYMA, where n is the number of the command; case is ignored.

(1) Assign expression to a name: (cn) *name* : *expression*;
    Example: (cn) `ex:(x+1)∧3+3*x-sin(x);`
    (Exponentiation is ∧)
    Forget assignment: (cn) `ex : "ex";`
(2) Comment: (cn) `"Quotes ($ prevents echo)."$`
(3) Constants: pi,e,i (cn) `%pi; %e; %i;`
(4) Define function f (in x): (cn) `f(x) :=` *expression*;
    Example: (cn) `f(x) := sin(x)*cos(x)+exp(x);`
    Note that $f(1)$ is valid.
(5) Denominator of ratio: (cn) `denom(a/b);`
(6) Differentiate: (cn) `diff(`*expression*`,`*variable*`);`
    Twice: (cn) `diff(`*expression*`,`*variable*`,2);`
    Several variables: (cn) `diff(`*expression*`,x,y);`
(7) Expand expression: (cn) `expand(`*expression*`);`
(8) Evaluate to number: (cn) `bfloat(`*expression*`);`

	To k decimal digits:	(cn) `fpprec:k; bfloat(`*expression*`);`
(9)	Factor expression:	(cn) `factor(`*expression*`);`
	Example:	(cn) `factor(x∧2+2*x+1);`
(10)	**For** loop:	(cn) `for` *var:val1* `thru` *val2* `do` *action*`;`
	Example: first 10 cubes	(cn) `for n:1 thru 10 do print(n∧3);`
(11)	Help:	(cn) `help();`
		(cn) `describe(`*command*`);`
		(cn) `apropos(`*word*`);`
(12)	**If** conditional:	(em) `if` *cond* `then` *true-action* `else` *false-action*`;`
	E.g., $f(x) = \begin{cases} 1 & \text{for } x < 1 \\ x & \text{for } x \geq 1 \end{cases}$	(cn) `f(x):= if x < 1 then 1 else x;`
(13)	Integrate:	(cn) `integrate(`*expression,variable*`);`
	Definite integral:	(cn) `integrate(`*exp,var,val1,val2*`);`
	Example, 0 to π:	(cn) `integrate(f(x),x,0,%pi):`
	Numerical approximation:	(cn) `romberg(f(x),x,0,%pi):`
(14)	Interrupt computation:	(cn) System interrupt key (`control-C`)
(15)	Limit:	(cn) `limit(`*expression,variable,value*`);`
	Example, at 0:	(cn) `limit(sin(x)/x,x,0);`
	Example, at infinity:	(cn) `limit(1/x,x,inf);`
(16)	Numerator of a ratio:	(cn) `numer(a/b);`
(17)	Part, second term in expr:	(cn) `part(ex,2);`
(18)	Plot:	(cn) `plot(`*expression,var,val1, val2*`);`
	Example, $-\pi$ to 1:	(cn) `plot(x+sin(x),x,-%pi,1);`
	Example, 3D:	(cn) `plot3d(x∧2+y∧2,x,-1,1,y,-2,1);`
(19)	Previous expression: %	(c2) `a+b∧2;`
		(cn) `integrate(%);`
	By number:	(cn) `integrate(c2);`
(20)	Read saved file:	(cn) `batch("`*filename*`");`
	MACSYMA lisp coded file:	(cn) `load("`*filename*`.1");`
(21)	Save expression in file:	(cn) `save("`*filename*`.1", ex);`
(22)	Simplify an expression:	(cn) `ratsimp(`*expression*`);`
(23)	Solve equation:	(cn) `solve(`*equation,variable*`);`
	Example:	(cn) `solve(x∧2+2*x+1=0,x);`
	System:	(cn) `solve((a*x+b*y=0,x+y=1),(x,y));`
(24)	Stop (end) session:	(cn) `quit();`
(25)	Substitute into expression:	(cn) `substitute(`*variable=value,expression*`);`
	Example:	(cn) `substitute(x=1,sin(x));`
(26)	Summation of expression:	(cn) `sum(`*expr,var,val1,val2*`);`
	Example, $\sum_{i=1}^{n} i^2$:	(cn) `sum(i∧2, i,1,n);`
(27)	Taylor series:	(cn) `taylor(`*expr,var,center,num-terms*`);`
	Example, at 0, degree 5:	(cn) `taylor(sin(x),x,0,5);`
	Several variables:	(cn) `taylor(sin(x+y),[x,y],[0,0],5);`
(28)	Show computation time:	(cn) `time:true;`
(29)	Truncate to an integer:	(cn) `Fix(`*real-number*`); "Actually Floor."$`
(30)	**While** loop:	(cn) `while` *condition* `do` *action-while-true*`;`

Example, while $n! \leq 10^{10}$ do (cn) n:1; "Initialize n";
compute $n!$, increment n: (cn) while n!<=10∧10 do (print(n!),n:n+1);

Example of Usage

Suppose we have the function $f : \Re \longrightarrow \Re$ defined by

$$f(x) = x^2 + \cos(x)$$

To define this function for MACSYMA, we enter

(cn) f(x) := x∧2 + cos(x) ;

Then we can evaluate f at 0 by

(cn) f(0);

or take the limit as x approaches 1 by

(cn) limit(f(x),x=1);

We can also give an expression $x^2 + \cos(x)$ a shorthand name by

(cn) ex : x∧2 + cos(x);

This is not a function, so evaluation ex(1) is not defined. But we can still use this expression in, e.g., finding the limit

(cn) limit(ex,x=1);

We can substitute 1 for x in expression ex by

(cn) substitute(x = 1, ex);

Finally, we can numerically evaluate symbolic expressions like $\sin(1)$ by

(cn) bfloat(sin(1));

A.4   Maple

Discussion

Maple is an interactive symbolic mathematics system. Among its capabilities are:

- Arbitrary precision arithmetic
- Calculus (differential, integral, and vector)
- Differential forms

- Expansion and factoring of expressions

- Graphing

- Group theory

- Infinite and finite series, limits, and products

- Linear algebra

- Number theory

- Set theory operations

- Solution of equations (non-linear, systems, and differential)

- Statistics

- Taylor's polynomials

Our description applies directly to version 4.2, but since only the basic features are discussed, it should also apply to later versions. For more information, see the on-line help (below) or the user's manual for Maple.

Getting Started

The following first steps are suggested:

(1) To start Maple enter `maple`. (Maple will read an initialization file "mapleinit" if you have one).

(2) To exit Maple enter `quit`.

Notation: In Maple functions are distinct from expressions. If `f` is a function, then evaluation produces an expression, `f(x)`. (Maple also has *operators* that transform functions into functions. For example, if `f` is a function, then `D(f)` is a new function, the derivative of `f`.)

Some Maple Commands

- indicates the primary prompt from Maple, and ≫ the secondary prompt.

(1)   Assign an expression to a name: ● `ex:=(x+1)∧3 + 3*x - sin(x);`
      (∧ or `**` is exponentiation)
      Note that *ex* is not a function,
            e.g., $ex(0) \neq 1$.
      Forget assignment: ● `ex := 'ex';`
(2)   Comment: ● `# Anything after an #.`
(3)   Constants: pi, e   ● `Pi; E;`
(4)   Define function *f*:   ● `f := proc(x)` *expression in var x* `end;`

Example:
- `f := proc(x) sin(x)*cos(x)+exp(x) end;`

Note that $f(1)$ is valid.

(5) Denominator of a ratio:
- `denom(a/b);`

(6) Differentiate:
- `diff(expression, variable(s));`

   Twice:
- `diff(expression,x$2);`

   Several variables:
- `diff(expression,x,y);`

(7) Expand expression:
- `expand(expression);`

(8) Evaluate to number:
- `evalf(expression);`

(9) Factor expression:
- `factor(expression);`

   Example:
- `factor(x∧2+2*x+1);`

(10) **For** loop:
- `for var from val1 to val2 do action od;`

   Example, first 10 cubes:
- `for n from 1 to 10 do n∧3 od;`

(11) **Help**:
- `help(topic);`

(12) **If** conditional:
- `If cond then true-action else false-action f`

E.g., $f(x) = \begin{cases} 1 & \text{for } x < 1 \\ x & \text{for } x \geq 1 \end{cases}$

- `f := proc(x) if x < 1 then 1 else x fi;`

(13) Integrate:
- `int(expression,variable);`

   Definite integral:
- `int(exp,variable=val1..val2);`

   Example, 0 to π:
- `int(f(x),x=0..Pi);`

   Numerical approximation:
- `evalf(int(f(x),x=0..Pi));`

(14) Interrupt computation:
System interrupt character, or button.

(15) Limit:
- `limit(expression,variable=value);`

   Example, at 0:
- `limit(sin(x)/x,x=0);`

   Example, at infinity:
- `limit(1/x,x=infinity);`

(16) Numerator of a ratio:
- `numer(a/b);`

(17) Part, second term in expression:
- `op(ex,2);`

(18) Plot:
- `plot(expression, var=val1..val2);`

   Example, $-\pi$ to 1:
- `plot(x+sin(x),x=-Pi..1);`

   Two functions:
- `plot({cos(x),sin(x)},x=-Pi..1);`

(19) Previous expression: "
- `a+b∧2;`
- `int(",x);`

(20) Read saved file:
- `read 'filename';`

   Maple coded file:
- `read 'filename.m';`

(21) Save expression in file:
- `save ex, 'filename.m';`

(22) Simplify an expression:
- `simplify(expression);`

(23) Solve equation:
- `solve(equation, variable);`

   Example:
- `solve(x∧2+2*x+1=0,x);`

   Linear system:
- `solve({a*x+b*y=0,x+y=1},{x,y});`

   Numerical:
- `fsolve(exp(x) + x = 0, x);;`

(24) Stop (end) session:
- `quit();`

(25) Substitute into expression:
- `subs(variable=value,expression);`

   Example:
- `subs(x=1, sin(x));`

(26) Sum an expression:
- `sum(expr, var=val1..val2);`

   Example, $\sum_{i=1}^{n} i^2$:
- `sum(i∧2, i=1..n);`

(27) Taylor series:
- `taylor(expr,var=center, deg error);`

Example, at 0, degree 8:   • `taylor(sin(x),x=0,9);`
Several variables: Do each variable separately.
Convert to polynomial: • `convert(taylor(sin(x),x=0,9),polynom);`

(28) Time elapsed since start: • `time();`
(29) Truncate to an integer: • `trunc(`*real number*`);`
(30) **While** loop: • `while` *condition* `do` *action-while-true* `od;`

Example, while $n! \leq 10^{10}$ do   • `n := 1; # Supply starting value.`
compute $n!$, increment n • `while n!<= 10^10 do n!; n := n+1 od;`

Example of Usage

Suppose we have the function $f : \Re \longrightarrow \Re$ defined by

$$f(x) = x^2 + \cos(x)$$

To define this function for Maple, we enter

• `f := proc(x) x^2 + cos(x) end;`

Then we can evaluate f at 0 by

• `f(0);`

or take the limit as x approaches 1 by

• `limit(f(x),x=1);`

We can also give an expression $x^2 + \cos(x)$ a shorthand name by

• `ex := x^2 + cos(x);`

This is not a function, so evaluation `ex(1)` is not defined. But we can still use this expression in, e.g., finding the limit

• `limit(ex,x=1);`

We can substitute 1 for x in expression `ex` by

• `subs(x = 1, ex);`

Finally, we can numerically evaluate symbolic expressions like $\sin(1)$ by

• `evalf(sin(1));`

A.5   Mathematica

Discussion

Mathematica is a computer algebra system designed for the interactive solution of mathematical problems. The capabilities of Mathematica include:

- Arbitrary precision arithmetic

- Calculus (differential, integral, and vector)

- Evaluation of special functions (Bessel, Gamma, etc.)

- Expansion and factoring of algebraic expressions

- Graphing (three dimensional and animation)

- Infinite and finite series, limits, and products

- Matrix and tensor algebra

- Numerical techniques

- Solutions of equations (systems, differential)

- Statistics

- Taylor's polynomials

One of Mathematica's strong points is its three-dimensional color graphics, which can be animated. (The Mathematica command

$$In[n]:=\ Plot3D[Exp[-(x\wedge2+y\wedge2)],\ \{x,-2,2\},\{y,-2,2\},Ticks\ ->None]$$

was used to produce the graphic on the cover.)

This description applies directly to version 1.2, but since only the basic features are discussed, it should also apply to later versions. Mathematica is divided into a "kernel" portion and a "front end" portion. The kernel does the computations and the front end handles interaction with the user. The kernel is the same for all computers; the user's interaction with the front end will vary between systems. Our description will be as general as possible, although any user should consult the manual for the front end used. Operations involving the computer system, such as starting, stopping, and file manipulation, will depend on the front end.

Getting Started

(1) Starting and exiting Mathematica depends on the user interface. There is a pause of up to a minute while the system starts.

(2) There is on-line help available, primarily concerned with the front end.

(3) Built-in functions have names beginning with a capital letter, e.g. `Sin`. (Mathematica is case sensitive; thus the user must take care in capitalization.) The command to integrate $\sin(x)\log(x)$ with respect to x would be entered: "`Integrate[Sin[x]*Log[x],x]`".

(4) Commands end by pressing the Enter key.

(5) High resolution graphics are available on many terminals, including the Macintosh.

(6) Input and output are numbered for later reference.

Notation: In Mathematica, function application uses brackets rather than parentheses. The names of built-in functions begin with an upper case letter. Thus to evaluate a (user-defined, not built-in) function f at 0 we enter `f[0]` and to differentiate it we enter `D[f[x],x]` (using the built-in function "`D`"). Curly braces, {}, are used for lists (where order counts) as well as for sets (where order does not count). Values are assigned using "=", except that ":=" is used for function definitions or delayed evaluation.

Table of Some Mathematica Commands

In[n] := indicates the nth input to Mathematica. On some Mathematica systems, there may be no prompt.

(1)	Assign expression to a name:	In[n]:= *name* = *expression*
	Example:	In[n]:= ex = (x+1)∧3 + 3*x - Sin[x]
	(∧ is exponentiation)	
	Forget assignment:	In[n]:= ex = .
(2)	Comment:	In[n]:= (* Enclose like this. *)
(3)	Constants: pi, e, i	In[n]:= Pi E I
(4)	Define function f (x variable):	In[n]:= f[x_] := *expression*
	Example:	In[n]:= f[x_] := Sin[x]*Cos[x]+Exp[x]
(5)	Denominator of ratio:	In[n]:= Denominator[a/b]
(6)	Differentiate:	In[n]:= D[*expression*, *variable*]
	Twice:	In[n]:= D[*expression*, {*variable*,2}]
	Several variables:	In[n]:= D[*expression*,x,y]
(7)	Expand expression:	In[n]:= Expand[*expression*]
(8)	Evaluate to number:	In[n]:= N[*expression*]
	To k decimal digits:	In[n]:= N[*expression*,k]
(9)	Factor expression:	In[n]:= Factor[*expression*]

Example: In[n]:= `Factor[x∧2+2*x+1]`

(10) **For** loop: In[n]:= `For[`*init, end test, next, action*`]`

Example, first 10 cubes: In[n]:= `For[n=1,n<=10,n=n+1,Print[n∧3]]`

(11) Help: In[n]:= `?`

(12) **If** conditional: In[n]:= `If[`*cond, true-action, false-action*`]`

E.g., $f(x) = \begin{cases} 1 & \text{for } x < 1 \\ x & \text{for } x \geq 1 \end{cases}$ In[n]:= `f[x_] := If[x<1,1,x]`

(13) Integrate: In[n]:= `Integrate[`*expression, variable*`]`

load "IntegralTables.m"   In[n]:= `<<IntegralTables.m`

Definite integral: In[n]:= `Integrate[`*exp, {variable, val1, val2}*`]`

Example, from 0 to π: In[n]:= `Integrate[f[x],{x,0,Pi}]`

Numerical approximation   In[n]:= `NIntegrate[f[x],{x,0,Pi}]`

(14) Interrupt computation: Control-C or Command-.

(15) Limit: In[n]:= `Limit[`*expression, variable->value*`]`

Example, at 0:   In[n]:= `Limit[Sin[x]/x,x -> 0]`

Example, at infinity: In[n]:= `Limit[1/x,x -> Infinity]`

(16) Numerator of ratio: In[n]:= `Numerator[a/b]`

(17) Part, second term in expression: In[n]:= `ex[[2]]`

(18) Plot: In[n]:= `Plot[`*expression, {var, val1, val2}*`]`

Example, from $-\pi$ to 1:   In[n]:= `Plot[x+Sin[x],{x,-Pi,1}]`

Two functions: In[n]:= `Plot[{Cos[x],Sin[x]},{x,-Pi,1}]`

(19) Previous expression: %   In[n]:= `a+b∧2`

     In[n+1]:= `Integrate[%,x]`

By number: In[n]:= `Integrate[In[m],x]`

(20) Read saved file: In[n]:= `<< "filename.m"`

(21) Save function "fun" in file: In[n]:= `fun >> "filename.m"`

(22) Simplify an expression: In[n]:= `Simplify[`*expression*`]`

(23) Solve equation: In[n]:= `Solve[`*equation, variable*`]`

Example: In[n]:= `Solve[x∧2+2*x+1 == 0,x]`

Linear system: In[n]:= `Solve[{x-y==0,x+y==1},{x,y}]`

Numerically: In[n]:= `FindRoot[Exp[x]+x==0,x,0]`

(24) Stop (end) session: Front End dependent

(25) Substitute into expression: In[n]:= *expr* `/.` *variable->value*

Example: In[n]:= `Sin[x] /. x->1`

(26) Sum: In[n]:= `Sum[`*expr, {var, val1, val2}*`]`

Example, $\sum_{i=1}^{n} i^2$:   In[n]:= `Sum[i∧2,{i,1,n}]`

(27) Taylor series: In[n]:= `Series[`*expr, {var, center, no. terms}*`]`

Example, at 0, degree 5:   In[n]:= `Series[Sin[x],{x,0,5}]`

Several variables: In[n]:= `Series[Sin[x+y],{x,0,5},{y,0,4}]`

Convert to polynomial: In[n]:= `Normal[%]`

(28) Time for computation: In[n]:= `Timing[`*command*`]`

(29) Truncate to an integer: In[n]:= `Floor[`*expression*`]` `(* No Trunc.*)`

(30) **While** loop: In[n]:= `While[`*condition, action-while-true*`]`

Example, while $n! \leq 10^{10}$ do   In[n]:= `n = 1 (* Supply starting value.*)`

compute $n!$, increment n: In[n]:= `While[n!<=10∧10,{n!,n=n+1}]`

Example of Usage

Suppose we have the function $f : \Re \longrightarrow \Re$ defined by

$$f(x) \ = \ x^2 \ + \ \cos(x)$$

To define this function in Mathematica we enter

     In[n]:= f[x_]:= x∧2+Cos[x]

(The underscore, _, indicates the variable on the left-hand side only.) Then we can evaluate f by

     In[n]:= f[0]

or take the limit as x approaches 1 by

     In[n]:= Limit[f[x],x -> 1]

We can also give an expression $x^2 \ + \ \cos(x)$ a shorthand name by

     In[n]:= ex = x∧2+Cos[x]

This is not a function, so evaluation ex[1] is not defined. But we can still use this expression in, e.g., finding the limit

     In[n]:= Limit[ex,x -> 1]

We can substitute 1 for x in expression ex by

     In[n]:= ex /. x->1

Finally, we can numerically evaluate symbolic expressions like sin(1) by

     In[n]:= N[Sin[1]]

A.6   muMATH

Discussion

MuMATH is a computer algebra system for microcomputers using the Apple II, MS-DOS, or CP/M-80 operating system. Since it runs on microcomputers, it is naturally less powerful than the systems for minicomputers or mainframes, such as Maple or MACSYMA. MuMATH can solve similar types of problems as the other systems, but the problems cannot be as difficult. Also, muMATH does not include graphing or pretty-printing, although some of these capabilities are available as extensions from third-party dealers. The partial pretty-printing, help, and graphics extensions from *CALC-87 muMATH Enhancements* are used in the table of commands below and are indicated by an asterisk.

    The capabilities of muMATH include:

- Arbitrary precision arithmetic

- Calculus (differential, integral, and vector)

- Expansion of algebraic expressions

- Graphing (with enhancement packages)

- Infinite and finite series and limits

- Matrix algebra

- Solutions of equations (systems, differential)

- Taylor's polynomials

This description applies directly to version 4.12 (with the CALC-87 enhancements) but should also apply to most versions.

MuMATH is no longer marketed for CP/M and PC-DOS systems, but an Apple II version is distributed and supported by Dr. Edwin Dickey, Educational Technology Center, College of Education, University of South Carolina, Columbia, SC 29208. In addition, an enhanced version of muMATH, renamed RIEMANN, is available for Atari ST computers (with other versions planned) from Begemann & Niemeyer Softwareentwicklung, Schwarzenbrinker Str. 91, 4930 DETMOLD, West Germany.

Getting Started

MuMATH is organized as a base system, muSIMP, and modules which are loaded into muSIMP to accomplish various tasks, such as differentiation, integration, etc. It is important to notice that muMATH uses only upper case; entries in lower case may not be understood. Thus, it is best to set the "Caps Lock" key when starting muMATH. To exit muMATH and return to the operating system, enter SYSTEM(); or a Control-C. There are demonstration programs which are run when a module is loaded. There are also a set of lessons for the use of muMATH as a calculator:

> CLES1.ALG rational arithmetic and assignment statements
> CLES2.ALG factorials and fractional powers
> CLES3.ALG polynomial expansion and factoring
> CLES4.ALG continued fractions and exponent simplifications

These can be run by loading them, e.g., "RDS('CLES1, 'ALG);".

Some muMATH Commands

? indicates the prompt from muMATH.

(1) Assign an expression to a name: ? *name* : *expression*
   Example: ? `EX: (X+1)∧3 + 3*X - SIN(X);`
   (∧ is exponentiation)
   Forget assignment: ? `EX:'EX;`
(2) Comment: ? `% After percent sign.`
(3) Constants: pi, e, i   ? `#PI #E #I;`
(4) Define function *F* (variable *x*): ? `FUNCTION F(X),` *expression* `ENDFUN;`
   Example: ? `FUNCTION F(X),`
                                        `SIN(X)*COS(X)+#E∧X ENDFUN;`
(5) Denominator of ratio: ? `DEN(A/B);`
(6) Differentiate: ? `DIF(`*expression*`,`*variable*`);`
   Twice: ? `DIF(`*expression*`,`*variable*`,2);`
   Several variables   ? `DIF(`*expression*`,X,Y);`
(7) Expand expression: ? `EXPAND(`*expression*`);`
(8) Evaluate to a number: Not directly available
(9) Factor expression: Not directly available
(10) **For** loop: ? *init*`; LOOP WHEN` *end-test* `EXIT,`
                                       action, *next* `ENDLOOP;`
   Example, first 10 cubes: ? `N:1; LOOP WHEN N>10 EXIT,`
                                       `PRINTLINE(N∧3),N:N+1 ENDLOOP;`
(11) Help: ? `HELP();`
(12) **If** conditional: ? `LOOP WHEN` *cond,* *true-action* `EXIT,`
                                          `WHEN 1=1,` *false-action* `EXIT,`
                                       `ENDLOOP;`
                                     `LOOP` not required when in `FUNCTION`

   E.g., $f(x) = \begin{cases} 1 & \text{for } x < 1 \\ x & \text{for } x \geq 1 \end{cases}$? `FUNCTION F(X),`

                                          `WHEN X < 1, 1 EXIT,`
                                          `WHEN X = 1, 1 EXIT,`
                                          `WHEN X > 1, X EXIT,`
                                       `ENDFUN;`
(13) Integrate: ? `INT(`*expression*`,`*variable*`);`
   Definite integral: ? `DEFINT(`*exp,variable,val1,val2*`);`
   Example, from 0 to π: ? `DEFINT(F(X),X,0,#PI);`
   Numerical approximation: Not available
(14) Interrupt computation: ? Press `ESCAPE-key`
(15) Limit: ? `LIM(`*expression, variable, value*`);`
   Example, at 0:   ? `LIM(SIN(X)/X,X,0);`
   Example, at positive infinity: ? `LIM(1/X,X,PINF);`
(16) Numerator of ratio: ? `NUM(A/B);`
(17) Part, second term in EX: ? `SECOND(EX); % EX considered as tree.`
(18) Plot: ? `GRAPH(`*expression,variable*`);`

(19) Previous expression: `@`

```
? A + B∧2;
? INT(@,X);
```

(20) Read saved file: `? RDS('NAME,'EXTN);`

(21) Save expressions in a file: Not directly available

Save entire environment: `? SAVE('`*filename*`);`

(22) Simplify an expression: `? TRGEXPD(`*expression*`);`

(23) Solve equations: `? SOLVE(`*equation*`,`*variable*`);`

Example: `? SOLVE(X∧2 + 2*X + 1 == 0,X);`

System (linear):   `? SOLVE([A*X+B*Y==0,X+Y==0],[X,Y]);`

(24) Stop (end) session: `? SYSTEM();`

(25) Substitute into equation: `? EVSUB(`*expression*`,`*variable*`,`*value*`);`

Example: `? EVSUB(SIN(X), X, 1);`

(26) Sum an expression: `? SUM(`*expression*`,`*variable*`,`*val1*`,`*val2*`);`

Example, $\sum_{i=1}^{n} i^2$:   `? SIGMA(I∧2, I, 1, N);`

(27) Taylor series: `? TAYLOR(`*expr*`,`*var*`,`*center*`,  `*degree*`);`

Example, at 0, degree 8:   `? TAYLOR(SIN(X), X, 0, 8);`

Several variables: Do each variable separately

(28) Time elapsed since reset: `? TIME();`

(29) Truncate to an integer: Not available

(30) **While** loop: `? LOOP WHEN `*cond*`, EXIT, `*action*` ENDLOOP;`

Example, while $n! \leq 10^{10}$ do compute $n!$, increment n

```
? N:1; % Supply starting value.
? LOOP WHEN N! > 10∧10 EXIT,
     PRINTLINE(N!),N:N+1,
  ENDLOOP;
```

Example of Usage

Suppose we have the function $f : \Re \to \Re$ defined by

$$f(x) = x^2 + \cos(x)$$

To define this function in muMATH, we enter

```
?    FUNCTION F(X), X∧2 + COS(X) ENDFUN;
```

Then we can evaluate `F` at 0 by

```
? F(0);
```

We can take the limit as x approaches 1 by

```
? LIM(F(X),X,1);
```

We can also give an expression $x^2 + \cos(x)$ a shorthand name by

```
? EX : X∧2 + COS(X);
```

This is not a function, so evaluation, `EX(1)`, is not defined. But we can still use this expression in, e.g., finding the limit

    ? LIM(EX,X,1);

We can substitute 1 for X in the expression EX by

    ? EVSUB(EX,X,1);

Evaluation of rationals to decimals can be done in muMATH by setting the variable POINT to the desired number of decimal points, e.g.,

    ? POINT: 10;

and then entering the rational expression. However, any constants must be assigned numerical values, e.g.,

    ? #PI : 3.14;

A.7   Theorist

Discussion

Theorist is a computer algebra system for the Macintosh. Among its capabilities are

- Expansion and factoring of algebraic expressions

- Solve equations and matrix operations

- Computation of limits and sums

- Calculus (differential and integral)

- Taylor polynomials

- Vector calculus

- 3D color graphics

Similar to Mathematica, Theorist has a "notebook" system for the presentation of results. The strong point of Theorist is the color graphics. For example, plotting $z = 1.3^x \sin(y)$ returns

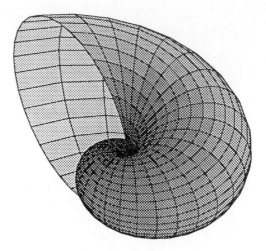

The plotting can be in color, the graphical objects can be easily manipulated with the mouse (including rotation of 3D objects and zoom), and the results can be saved in PostScript files for use in other documents.

Theorist is a less powerful algebra system than the others. For example, although Theorist integrates some functions without help, integration by parts requires the user to pick out the parts to be used. A table of integrals can be loaded to help simplify difficult integrals.

Theorist does not compute limits, has at most 19 digits of precision, and has no programming language.

Usage

The user can enter an expression by using the keyboard or by selecting symbols from a palette. Several operators, such as simplification, integration, Taylor series, evaluation, graphing, etc., can then be applied to the expression. The user can define a function in terms of the usual collection of built-in functions.

Bibliography

[1] Akritas, A., *Elements of Computer Algebra with Applications*, Wiley, New York, 1989.

[2] Char, B. W., K. O. Geddes, G. H. Gonnet, and S. M. Watt, *Maple User's Guide*, Watcom Publications, Ontario, Canada, 1985.

[3] Crilly, T., "From Fixed Points to Continued Fractions," *The Mathematical Gazette*, **73**(463), March 1989, p. 16.

[4] Davenport, J. H., Y. Sivet, and E. Tournier, *Computer Algebra*, Academic Press, San Diego, 1988.

[5] Herman, E. A., "Derive, A Mathematical Assistant, ver 1.22," *The American Mathematical Monthly*, **96**(10), December 1989, p. 948.

[6] Hosack, J. M., "A Guide to Computer Algebra Systems," *College Mathematics Journal*, **17**(5), November 1986, p. 434.

[7] Kroll, L. S., "Mathematica–A System for Doing Mathematics by Computer Algebra," *The American Mathematical Monthly*, **96**(9), November 1989, p. 855.

[8] Norman, A. C., "Algebraic Manipulation," *Encyclopedia of Computer Science and Engineering*, Van Nostrand Reinhold, New York, 1983, p. 41.

[9] Pavelle, R., M. Rothstein, and J. Fitch, "Computer Algebra," *Scientific American*, **245**(6) , December 1981, p. 102.

[10] Small, D. and J. Hosack, "Computer Algebra Systems: Tools for Reforming Calculus," *Toward a Lean and Lively Calculus*, Mathematical Association of America Notes Number 6, 1986, p. 143.

[11] ———, J. Hosack, and K. Lane, "Computer Algebra Systems in Undergraduate Instruction," *College Mathematics Journal*, **17**(5), November 1986, p. 423.

[12] Wilf, H. S., "The Disk with the College Education," *American Mathematical Monthly*, **89**, January 1982, p. 4.

[13] Wolfram, S., "Computer Software in Science and Mathematics," *Scientific American*, **251**, September 1984, p. 188.

[14] ———, *Mathematica–A System for Doing Mathematics by Computer*, Addison-Wesley, Reading, Mass.. 1988.

[15] Yun, D. Y. Y. and D. R. Stoutemyer, "Symbolic Mathematical Computation," *Encyclopedia of Computer Science and Technology*, **15**, M. Dekker, New York, 1980, p. 235.

Index